METHODS IN MOLECULAR BIOLOGY™

Series Editor
John M. Walker
School of Life Sciences
University of Hertfordshire
Hatfield, Hertfordshire, AL10 9AB, UK

For further volumes:
http://www.springer.com/series/7651

The Low Molecular Weight Proteome

Methods and Protocols

Edited by

Helena Bäckvall and Janne Lehtiö

Science for Life Laboratory, Department of Oncology-Pathology, Karolinska Institutet, Stockholm, Sweden

Editors
Helena Bäckvall
Science for Life Laboratory
Department of Oncology-Pathology
Karolinska Institutet
Stockholm, Sweden

Janne Lehtiö
Science for Life Laboratory
Department of Oncology-Pathology
Karolinska Institutet
Stockholm, Sweden

ISSN 1064-3745 ISSN 1940-6029 (electronic)
ISBN 978-1-4614-7167-7 ISBN 978-1-4614-7209-4 (eBook)
DOI 10.1007/978-1-4614-7209-4
Springer New York Heidelberg Dordrecht London

Library of Congress Control Number: 2013937719

Printed on acid-free paper

Humana Press is a brand of Springer
Springer is part of Springer Science+Business Media (www.springer.com)

Preface

The aim of this book is to provide a useful resource for experienced proteomics practitioners as well as an aid to newcomers who wish to become acquainted with the theory and practice of a wide array of methods in analyzing small proteins or peptides. Small proteins with molecular weights of <25 kDa are involved in major biological processes such as ribosome formation, stress adaption, cellular signaling, and cell cycle control. Fifty percent of the human proteins are below 26.5 kDa (Fig. 1) illustrating the importance of efficient methods for analysis of low molecular weight proteome. The distinction between peptide and small protein is vague, hence we have included in this book methods suitable for both peptide and protein analysis with focus on methods and application that can be used for small protein analysis. The study of the low molecular weight proteome has identified many central regulators of biology such as cytokines, chemokines, peptide hormones, and proteolytic fragments of larger proteins. BRCA2, a cancer-related low-abundant protein, is found in the serum as fragments bound to albumin, highlighting the importance of the low molecular weight proteome in biomarker discovery. Another example is the largest calcium-binding protein family, the S100-protein family, where all known members have molecular weights around 10 kDa. Despite their importance, the coverage of smaller proteins in standard proteome studies is rather sparse. The underrepresentation of low molecular weight proteins may be attributed to the low numbers of proteolytic peptides formed by tryptic digestion as well as their tendency to be lost in protein separation and concentration/desalting procedures. This group of proteins and peptides also migrate out of gels easily as well as stain poorly,

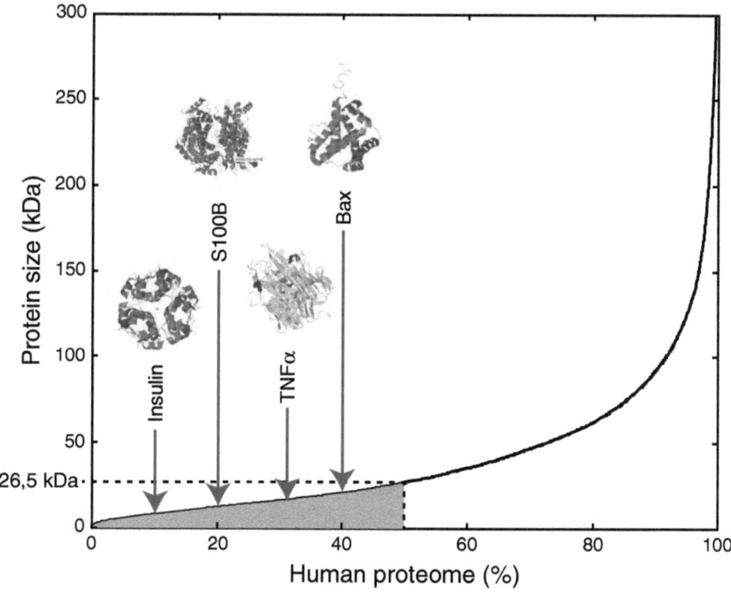

Fig. 1 Distribution of molecular weights of proteins in the human proteome. Fifty percent of all proteins have a calculated molecular weight below 26.5 kDa

lowering the detection sensitivity in gel-based studies. There are various tailor-made strategies that can be used for low molecular weight proteome analyses, which vary in their aims and the technical equipment. Due to the unique features of these proteins, the technical challenges differ somewhat from those in "common" proteomics.

In this volume of *Methods in Molecular Biology*, we provide protocols for analysis of low molecular weight proteins and peptides, protocols for such methods applied in clinical research and an up-to-date review of quantitative protein profiling by labeling. The book is divided into three parts: (1) *Analysis methods for low molecular weight proteins and peptides*, with eight method chapters providing protocols to perform prefractionation to reduce complexity in samples, depletion strategy for removal of high-abundant proteins, labeling for protein profiling and quantification, how to develop and use an MRM assay for targeted detection of peptides, material-enhanced laser desorption/ionization (MELDI) a new form of the laser desorption/ionization technique to profile low molecular weight proteome, how to perform immunoprecipitation on peptide levels followed by MALDI (iMALDI), localizing small ionizable peptides and proteins in tissue sections with MALDI imaging, and the multiplexed antibody suspension bead arrays which can function as validation assays or affinity proteomics profiling approach; (2) *Bioinformatics*, a chapter describing how to improve quantitative accuracy and precision in MS-based proteomics by PQPQ-algorithm; (3) *Applications*: How to collect and prepare samples for peptidomics, a method for screening large numbers of urine samples, in vitro phage display selection of peptide ligands in fractions of rat kidney, a description of a methods for protein expression profiling on brain tumor samples using SELDI-MS, and a workflow of MHC ligand identification in clinical materials.

In keeping with the established format of the *Methods in Molecular Biology* series, each chapter begins with a description of the basic theory behind the method being described. The Materials section lists all the chemicals, reagents, buffers, and other materials necessary for carrying out the method. The Notes section complements the Methods and Material section by indicating how to choose good products and how best to deal with any problem or difficulties that may arise when using the technique.

Experienced scientists have contributed to this volume, which is intended to give an overview of the contemporary challenges and possibilities in the various areas of peptidomics and to offer some detailed protocols as examples for successful analysis in applied proteomics studies. It is likely that many possible major findings reside in this understudied group of proteins and possibly many short open reading frames are still unannotated even in well-studied organisms. Therefore, we hope that this book can raise your interest in the low molecular weight proteome analysis and that this book will be a valuable reference book for your laboratory work.

Stockholm, Sweden

Helena Bäckvall
Janne Lehtiö

Acknowledgements

The editors would like to acknowledge and thank Dr. Lukas Orre and Ph.D. student Yafeng Zhu for helping us to create Fig. 1 illustrating the low molecular weight proteome.

Contents

Part III Methods Applied in Clinical Research

Contributors

YANMING AN • *Lombardi Comprehensive Cancer Center, Georgetown University, Washington, DC, USA*

MALIN ANDERSSON • *Department of Pharmaceutical Bioscience, Drug safety and Toxicology, Uppsala University, Uppsala, Sweden*

CRINA I.A. BALOG • *Department of Parasitology, Leiden University Medical Center, RC Leiden, The Netherlands*

JONAS BERGQUIST • *Department of Physical and Analytical Chemistry, Uppsala University, Uppsala, Sweden*

GÜNTHER K. BONN • *Institute of Analytical Chemistry and Radiochemistry, Leopold-Franzens University Innsbruck, Innsbruck, Austria; Sino-Austrian Center for Biomarker Research, Peking University–Health Science Center, Beijing, China*

CHRISTOPH H. BORCHERS • *University of Victoria-Genome British Columbia Proteomics Centre, University of Victoria, Victoria, BC, Canada; Department of Biochemistry and Microbiology, University of Victoria, Victoria, BC, Canada*

ANDRÉ M. DEELDER • *Department of Parasitology, Leiden University Medical Center, RC Leiden, The Netherlands*

RICO DERKS • *Department of Parasitology, Leiden University Medical Center, RC Leiden, The Netherlands*

DOMINIK DOMANSKI • *University of Victoria-Genome British Columbia Proteomics Centre, University of Victoria, Victoria, BC, Canada*

KIMI DROBIN • *Science for Life Laboratory Stockholm, School of Biotechnology, KTH - Royal Institute of Technology, Stockholm, Sweden*

JENNY FORSHED • *Science for Life Laboratory, Department of Oncology–Pathology, Karolinska Institutet, Stockholm, Sweden*

RADOSLAV GOLDMAN • *Lombardi Comprehensive Cancer Center, Georgetown University, Washington, DC, USA*

JÖRG HANRIEDER • *Department of Physical and Analytical Chemistry, Uppsala University, Uppsala, Sweden*

RUDIGER HESS • *PXBioVisioN GmbH, Hannover, Germany*

ANNA LJUNGDAHL • *Department of Pharmaceutical Bioscience, Drug safety and Toxicology, Uppsala University, Uppsala, Sweden*

KIE KASUGA • *Science for Life Laboratory, Department of Oncology–Pathology, Karolinska Institutet, Stockholm, Sweden*

MICHAEL A. KUZYK • *University of Victoria-Genome British Columbia Proteomics Centre, University of Victoria, Victoria, BC, Canada*

TAE-HWAN KWON • *Department of Biochemistry and Cell Biology, School of Medicine, Kyungpook National University, Taegu, South Korea*

BYUNG-HEON LEE • *Department of Biochemistry and Cell Biology, School of Medicine, Kyungpook National University, Taegu, South Korea*

JOHAN LENGQVIST • *Biopharmaceutical Research Unit, Department of Protein Science, Novo Nordisk A/S, Måløv, Denmark; Proteomics Core Faciltiy, Sahlgrenska Academy, Gothenburg University, Gothenburg, Sweden*
OLEG A. MAYBORODA • *Department of Parasitology, Leiden University Medical Center, RC Leiden, The Netherlands*
PETER NILSSON • *Science for Life Laboratory Stockholm, School of Biotechnology, KTH - Royal Institute of Technology, Stockholm, Sweden*
CAROL E. PARKER • *University of Victoria-Genome British Columbia Proteomics Centre, University of Victoria, Victoria, BC, Canada*
MARIA PERNEMALM • *Science for Life Laboratory, Department of Oncology–Pathology, Karolinska Institutet, Stockholm, Sweden*
MATTHIAS RAINER • *Institute of Analytical Chemistry and Radiochemistry, Leopold-Franzens University Innsbruck, Innsbruck, Austria; Sino-Austrian Center for Biomarker Research, Peking University–Health Science Center, Beijing, China*
JENNIFER D. REID • *University of Victoria-Genome British Columbia Proteomics Centre, University of Victoria, Victoria, BC, Canada*
CONSTANTIN SAJDIK • *Institute of Analytical Chemistry and Radiochemistry, Leopold-Franzens University Innsbruck, Innsbruck, Austria; Sino-Austrian Center for Biomarker Research, Peking University–Health Science Center, Beijing, China*
ANNSOFI SANDBERG • *Science for Life Laboratory, Department of Oncology–Pathology, Karolinska Institutet, Stockholm, Sweden*
JOCHEN M. SCHWENK • *Science for Life Laboratory Stockholm, School of Biotechnology, KTH - Royal Institute of Technology, Stockholm, Sweden*
BRINDA SHAH • *University of Victoria-Genome British Columbia Proteomics Centre, University of Victoria, Victoria, BC, Canada*
HARALD TAMMEN • *PXBioVisioN GmbH, Hannover, Germany*
CARL WIBOM • *Department of Oncology, Institution for Radiation Sciences, Umeå, Sweden*

Part I

Low Molecular Weight Analysis Methods

Chapter 1

Narrow-Range Peptide Isoelectric Focusing as Peptide Prefractionation Method Prior to Tandem Mass Spectrometry Analysis

Maria Pernemalm

Abstract

High sample complexity is one of the major challenges in mass spectrometry-based proteomics today. Despite massive improvement in instrumentation, sample prefractionation is still needed to reduce sample complexity and improve proteome coverage. Isoelectric focusing (IEF) has been traditionally used as a first-dimension protein separation technique in two-dimensional gel electrophoresis-based proteomics. Recently, *peptide* IEF has emerged as appealing alternative for anion exchange chromatography in multidimensional LC-MS/MS workflows. The rationale behind using narrow-range peptide isoelectric focusing as a prefractionation method prior to ms/ms is to reduce the complexity induced by tryptic digestion. This is done by selectively analyzing a sub-fraction of peptides with an acidic p*I*. The p*I* range is chosen as it has previously been shown that 96 % of human proteins have at least one tryptic peptide between pH 3.4 and 4.9. This ensures high proteome coverage while reducing the number of peptides with 2/3. In addition the focusing precision is optimal in this range. Therefore, by analyzing this sub-fraction of peptides the complexity of the sample can be reduced without significant loss of proteome coverage. As the theoretical p*I* of peptides can be calculated, the p*I* of the identified peptides can be used to validate the peptide sequence (identified peptides with p*I* outside the pH range 3.4–4.9 are more likely to be false positives). In addition, this approach is compatible with iTRAQ labelling as the different iTRAQ labels migrate similarly in IEF.

Key words Narrow-range peptide isoelectric focusing, Prefractionation, Mass spectrometry, Proteomics, p*I*, iTRAQ

1 Introduction

Technical differences between individual mass spectrometers related to sensitivity and mass accuracy greatly influence the performance of proteomics analyses.

In addition, the level of sample complexity influences the performance of the mass spectrometry analysis. High sample complexity in proteomics samples is characterized by large number of chemically diverse analytes and a high dynamic range of concentrations.

Helena Bäckvall and Janne Lehtiö (eds.), *The Low Molecular Weight Proteome: Methods and Protocols*, Methods in Molecular Biology, vol. 1023, DOI 10.1007/978-1-4614-7209-4_1,

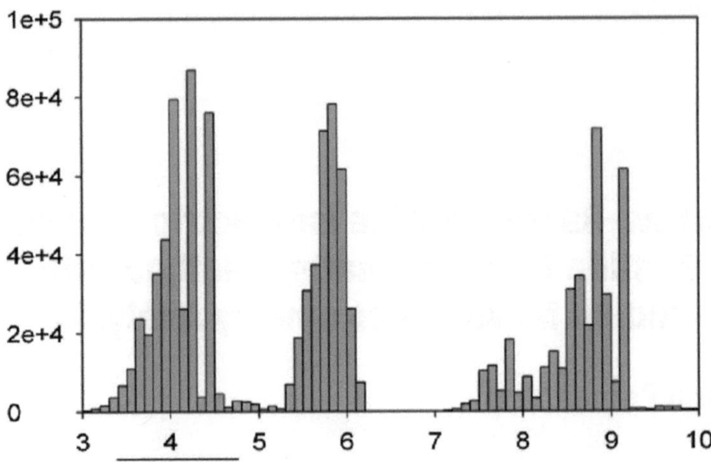

Fig. 1 Human tryptic peptide p*I* distribution. A plot of the predicted p*I* values for all human tryptic peptides, based on database consisting of all translated human sequences from ENSEMBL. All unmodified peptides with 4–60 amino acids and no missed cleavages are included. Approximately one third is in the pH interval 3.4–4.9, indicated by a *black bar*. Number of peptides shown on y-axis and peptide pI on x-axis

These sample characteristics are of analytical importance as they are influenced by technical limitations in mass spectrometry.

To overcome these analytical challenges the most common approach is to reduce the sample complexity by prefractionation. Prefractionation can be performed either on a protein level, or on a peptide level, or using a combination of the two. The prefractionation method described in this chapter, *narrow-range peptide isoelectric focusing*, is targeting the complexity induced by tryptic digestion [1–3](Fig. 1) and has successfully been applied in combination with several upstream fractionation methods, such as high abundant protein depletion in plasma [3], membrane protein enrichment from cell lines [4], and mouse liver proteome [5].

In theory, any complex protein sample could be applied to tryptic digestion and separation on immobilized pH strips (IPG strips) as described below. The acetone precipitation is performed to ensure sample buffer compatibility with the downstream analysis, but could be replaced by buffer exchange and concentration on cutoff filters (e.g., 3 kDa cutoff, Millipore). The digestion protocol is an example of a protocol compatible with iTRAQ labelling [6] and could be replaced with any other standard tryptic digestion protocol. iTRAQ labelling is performed to enable relative quantification between samples; however the iTRAQ labelling step is optional and could be excluded. The protocol should also be compatible with isotopic labelling such as SILAC labelling, although this has not been reported in the literature. The SCX cleanup has several benefits in this protocol; it removes SDS as well as excess iTRAQ label and in addition desalts the sample and changes the buffer to a volatile buffer.

2 Materials

2.1 Precipitation of Proteins

1. Ice-cold acetone.

2.2 Tryptic Digestion

1. Water purified with a Milli-Q purification system (Millipore, Bedford, MA).
2. Dissolution buffer: 0.5 M triethylammonium bicarbonate (TEAB) in Milli-Q water.
3. Detergent: 2 % SDS in Milli-Q water.
4. Reducing agent: 50 mM Tris-(2-carboxyethyl)-phosphine (TCEP) in Milli-Q water.
5. Cysteine blocking agent: 200 mM Methyl methanethiosulfonate (MMTS) in isopropanol.
6. Enzyme: Trypsin (sequence grade modified trypsin, Promega), 0.2 μg/μl in Milli-Q water.

2.3 8-plex iTRAQ Labelling

1. 8-plex iTRAQ kit (Applied Biosystems).
2. Isopropanol.

2.4 SCX Cleanup

1. SCX cartridges (Strata X-C, 30 mg/1 ml, Phenomenex).
2. Wash solution 1: 100 % Methanol.
3. Wash solution 2: Milli-Q water.
4. Formic acid.
5. Wash solution 3: 30 % methanol, 0.1 % formic acid.
6. Elution buffer: 30 % methanol, 5 % ammonium hydroxide.

2.5 Narrow-Range IEF

1. IPG strips 24 cm linear gradient pH 3.5–4.5 (GE Healthcare).
2. Rehydration solution: 8 M Urea with bromophenol blue (BPB) (i.e., 2 mg BPB in 10 ml 8 M Urea).
3. Rehydration solution with Pharmalyte: 8 M Urea with bromophenol blue, 1 % Pharmalyte pH 2.5–5 (GE Healthcare).
4. Oil: Dry Strip Cover Fluid, Plus one (GE Healthcare).
5. Electrode paper, Wicks (GE Healthcare).
6. Sample cups (GE Healthcare).

2.6 Peptide Elusion

1. Water purified with Milli-Q purification system (Millipore, Bedford, MA).
2. Sharp, clean, scissors.
3. Forceps.

3 Methods

The sample to start with in this protocol should preferably be a protein sample with known protein concentration. As a quality control an SDS page gel could be run and stained prior to tryptic digestion. An aliquot of digested sample could then be run on SDS page gel after tryptic digestion to make sure that all proteins are digested into peptides. If iTRAQ labelling is performed, the sample should contain no buffers or additives containing free primary amines (e.g., Tris, ammonium bicarbonate), as the iTRAQ label reacts with primary amines.

Please note that the sample preparation protocol has two overnight steps (digestion and IEF preparation) and that the isoelectric focusing itself can take over 24 h, and plan ahead for this. The protocol can be stopped at several stages and these will be indicated in the text.

3.1 Protein Precipitation

Protein precipitation is performed in order to concentrate the sample and change buffers prior to digestion. In order to obtain as good yield as possible the protein concentration in your sample should not be less than 1 mg/ml (*see* **Note 1**).

1. Take out 100 μg of each sample and add it to an Eppendorf tube (*see* **Note 2**).
2. Add four volumes of ice-cold acetone.
3. Invert the tube a couple of times.
4. Incubate the sample 1 h on ice, or until a flocculent forms.
5. Centrifuge the sample at 10,000 × *g* at +4 °C.
6. Remove the liquid with a pipette, be careful not to touch the pellet.
7. Allow the remaining acetone to vaporize. Do not over dry the pellet, as this may make it difficult to resuspend.

3.2 Tryptic Digestion

Tryptic digestion is performed prior to mass spectrometry in order to perform efficient amino acid sequence determination. The protocol below is compatible with iTRAQ labelling as it contains no free amines. It could, however, be applied even if not performing iTRAQ labelling. Similarly, other protocols for tryptic digestion could be used in this step if preferred.

1. Add 20 μl of 0.5 M TEAB to each sample.
2. Add 1 μl of 2 % SDS to solubilize the pellet, vortex (*see* **Note 3**).
3. Add 2 μl of 50 mM TCEP, vortex.
4. Incubate the sample for 1 h at 60 °C.
5. Add 1 μl of 200 mM MMTS (*see* **Note 4**).
6. Incubate at room temperature for 10 min.
7. Dissolve 2 μg of trypsin in 10 μl Milli-Q water per sample.

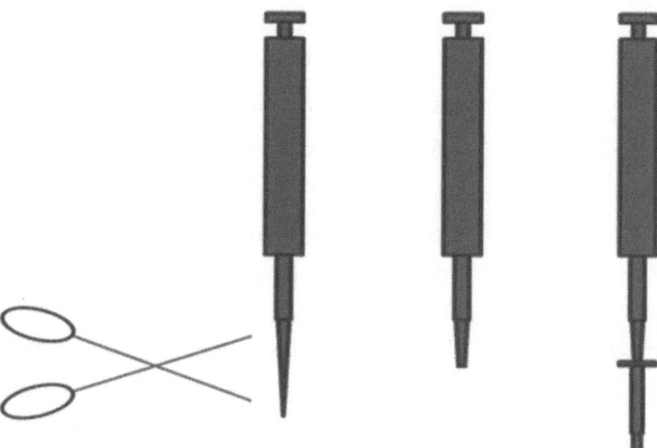

Fig. 2 SCX setup. First apply the fluid to the cartridge. Then change tip. Use normal disposable 1 ml plastic tips. Cut the tip in half (approximately, depends on tip shape and brand) and put it on to a 1 ml pipette. Make sure that the tip fits into the cartridge without touching the fluid. Set the pipette to 1 ml and push the fluid through the cartridge by pushing air through the pipette. Repeat the procedure if not all liquid is pushed through the cartridge the first time

8. Add the trypsin to the sample and digest over night at 37 °C (ratio 1:50, trypsin:total protein) (*see* **Note 5**).

3.3 8-plex iTRAQ Labelling (See Note 6)

The following protocol is adapted from the iTRAQ manufacturer (Applied Biosystems).

1. Allow each vial of iTRAQ label to reach room temperature.
2. Spin the solution to the bottom of the vial.
3. Add 50 μl of isopropanol to each iTRAQ vial.
4. Add the content of one iTRAQ label vial to each of the digested samples, remember to note which label is added to what sample.
5. Vortex the samples.
6. Incubate at room temperature for 2 h.
7. Combine the eight labelled samples into one tube.

3.4 SCX Cleanup

In order to remove excess iTRAQ label and to desalt and concentrate the peptide sample prior to the isoelectric focusing an SCX cartridge is used. To maximize the peptide yield, the SCX cartridge is a mixed-mode cartridge, with both SCX and reversed phase characteristics. The SCX cleanup described below is preferably performed, even if not using iTRAQ labelling, to remove salts and buffers prior to isoelectric focusing. The cartridges can be eluted either with a vacuum manifold or by applying manual pressure using a 1 ml pipette as shown below (Fig. 2).

1. Wash the column with 600 μl of 100 % methanol.
2. Wash the column with 600 μl of Milli-Q water.

3. Adjust the sample volume to minimum 600 μl and acidify to end concentration of 0.1 % formic acid.
4. Apply the sample to the column.
5. Wash away unbound material with 600 μl of wash solution 3 (30 % methanol, 0.1 % formic acid).
6. Elute the sample in 600 μl elution buffer.
7. Dry the sample in a speed vac system.

3.5 Narrow-Range IEF

This protocol assumes the use of IPGphor isoelectric focusing system (GE Healthcare) or equivalent equipment for isoelectric focusing. The isoelectric focusing in this protocol is performed on normal commercially available IPG strips in the pH range 3.5–4.5 (GE Healthcare). It is important that the peptide sample contains no salts, as this will interfere with the focusing (*see* **Note 7**). Once the focusing is done it is very important to start the cutting and elusion directly as the peptides start to diffuse within the gel (*see* **Note 8**).

Day 1

1. Prepare rehydration solution. Extra rehydration solution can be frozen and used later.
2. Prepare rehydration solution with 1 % pharmalyte pH 2.5–5. 250 μl per strip will be needed.
3. Distribute 250 μl of rehydration solution with 1 % pharmalyte pH 2.5–5 in as many lanes of the rehydration tray as you have strips.
4. Take off the plastic sealing on the gel strips (pH 3.5–4.5) and place the strip, gel-side down (you can read the text), on the solution. Make sure the whole strip soaks the solution.
5. Add a few drops of rehydration solution in the surrounding empty lanes to avoid drying of the strip.
6. Put an airtight plastic bag/cover over the tray and leave overnight in room temperature to swell.

Day 2

1. Dissolve the sample in as little volume of rehydration solution with pharmalyte as possible, i.e., 150 μl of rehydration solution with 1 % pharmalyte pH 2.5–5.
2. Place the manifold (focusing tray) in the IPGphor.
3. Transfer the strips to the manifold with gel side up and +-mark upwards towards the anode. The upper edge of the gel should be approximately on the first dot (under the numbers) in the manifold.

4. Wet the electrode papers with 150 μl of Milli-Q water. They should be damp, but not dripping wet. Press with your finger (wearing gloves) to remove excess water.
5. Put the papers at both cathode and anode side of the strip.
6. Put on the electrodes.
7. Position the cup close to the cathode end of the strip.
8. Load the sample in the cup.
9. Fill the manifold with approximately 135 ml of dry strip cover fluid; make sure that the strips are covered with oil (*see* **Note 9**).
10. Start the run with the following program:

    Step 1: 500 V, 1 h gradient.

    Step 2: 2,000 V, 2 h gradient.

    Step 3: 4,000 V, 2 h gradient.

    Step 4: 6,000 V, 2 h gradient.

    Step 5: 8,000 V, 2 h gradient.

    Step 6: 8,000 V, hold.
11. Let the run continue until at least 100 kVh is reached. This takes about 24 h. Change filter paper if the current is too high (maximum 50 μA/strip, this can be increased to 75 μA/strip).

Day 3

1. Remove as much dry strip cover fluid as possible by rinsing with Milli-Q water. Immerse the strips repeatedly in a water bath and dry the plastic side of the strip gently on paper napkin.
2. Proceed immediately to peptide elution/extraction step.

3.6 Peptide Elution/Extraction

1. As soon as possible after the focusing has been stopped, cut the strip in as many pieces you think is appropriate, e.g., 24 pieces á 1 cm. The easiest way to do it is to use sharp, clean, scissors and holding the strip in the + end (with your gloved fingers or forceps, be careful not to touch the gel), cut the pieces directly into Eppendorf tubes (*see* **Note 10**). The pieces can now be frozen until extraction.
2. Extraction is performed by adding 50–100 μl of Milli-Q water to the pieces (make sure the pieces are covered in water). Let stand on moderate shaking in room temperature for 1 h.
3. Remove the supernatant.
4. Repeat two times (three extractions in total).
5. Dry the samples, e.g., in a speed vac.
6. Put in −80 °C until mass spectrometry analysis (*see* **Note 11**).

4 Notes

1. If the concentration is less than 1 mg/ml buffer exchange and sample concentration on 3 kDa cutoff filters (Millipore) is an option. The drawback with using cutoff filter is that it is very time consuming and introduces sample loss and additional variability.
2. One hundred micrograms is the starting amount in this protocol, as it is a suitable amount to run on IEF strips even if not doing iTRAQ labelling. Loading less than 100 μg on the strip could reduce the sensitivity of the method.
3. If the solution is not completely clear, because the pellet is not completely solubilized, it often is solubilized during the digestion. If the pellet is very difficult to dissolve, heating at 70 °C, sonicating in sonicator bath, or adding urea (<1 M end concentration) can be optional strategies. If the sample is not soluble post precipitation buffer exchange and sample concentration on 3 kDa cutoff filters (Millipore) is an option.
4. When searching the ms/ms data do not forget to specify that MMTS has been used instead of iodoacetamide, which is usually the standard setting.
5. To ensure that the digestion has worked you can take off an aliquot of each sample before digestion and run on SDS page and stain with silver (e.g., 1 μg). Take off the same amount after digestion and run another gel; make sure that no bands are present above 10 kDa. Another option is to filter the digest through a 10 kDa cutoff membrane to remove undigested proteins.
6. If you want to analyze more than eight samples a pooled internal standard can be used as common sample between different iTRAQ pools. The pooled internal standard is preferably made up by pooling aliquots from each of the tryptic digests in the study. By doing this all peptides in the individual samples should be represented in the pooled internal standard. The pooled standard is then used to normalize all other samples in the pool against.
7. When preparing your sample for IEF, try keeping the salt concentrations as low as possible, preferably <10 mM. This is to avoid too high currents in the run.
8. The peptides will start to diffuse within the strip as soon as you switch off the focusing. To maintain focusing precision it is very important to cut the IPG strip and elute the peptides immediately after stopping the focusing. Handle strips with care, always use gloves, and do not touch the gel side.

9. During the focusing, it is very important that the strips are covered with oil; otherwise the urea will start to precipitate.
10. Try to cut the samples at a 90° angle, to reduce the spread of peptides over several fractions. When cutting manually it can be quite difficult to cut even-sized pieces, so therefore it could be an advantage to analyze continuous fractions at the mass spectrometry stage to make sure that the same population of peptides is sampled between different IPG strips.
11. The theoretical p*I* of each of the identified peptides (only non-modified peptides) can be calculated and used to confirm the identification. If the peptide has a p*I* outside the pH range 3.5–4.5 it is more likely to be a false positive. In addition the theoretical p*I* of the peptide can be used to calculate the focusing accuracy of the IEF. Peptide spread over fractions can also be good to evaluate to decide the most suitable fraction width.

References

1. Cargile BJ, Bundy JL, Freeman TW, Stephenson JL Jr (2004) Gel based isoelectric focusing of peptides and the utility of isoelectric point in protein identification. J Proteome Res 3: 112–119
2. Essader AS, Cargile BJ, Bundy JL, Stephenson JL Jr (2005) A comparison of immobilized pH gradient isoelectric focusing and strong-cation-exchange chromatography as a first dimension in shotgun proteomics. Proteomics 5:24–34
3. Eriksson H, Lengqvist J, Hedlund J, Uhlen K, Orre LM, Bjellqvist B, Persson B, Lehtio J, Jakobsson PJ (2008) Quantitative membrane proteomics applying narrow range peptide isoelectric focusing for studies of small cell lung cancer resistance mechanisms. Proteomics 8:3008–3018
4. Pernemalm M, De Petris L, Eriksson H, Branden E, Koyi H, Kanter L, Lewensohn R, Lehtio J (2009) Use of narrow-range peptide IEF to improve detection of lung adenocarcinoma markers in plasma and pleural effusion. Proteomics 9:3414–3424
5. Lee A, Chick JM, Kolarich D, Haynes PA, Robertson GR, Tsoli M, Jankova L, Clarke SJ, Packer NH, Baker MS (2011) Liver membrane proteome glycosylation changes in mice bearing an extra-hepatic tumour. Mol Cell Proteomics 10(9):M900538MCP200
6. Lengqvist J, Uhlen K, Lehtio J (2007) iTRAQ compatibility of peptide immobilized pH gradient isoelectric focusing. Proteomics 7: 1746–1752

Chapter 2

Analysis of Peptides by Denaturing Ultrafiltration and LC-MALDI-TOF-MS

Yanming An and Radoslav Goldman

Abstract

The dynamic range of complex biological samples represents a challenge for mass spectrometric characterization. Removal of high abundant proteins is a prerequisite for a successful mass spectrometric analysis of low abundant analytes. In particular, plasma and serum proteome span at least ten orders of magnitude and represent a major challenge for biomarker discovery. Immunoaffinity depletion is the most common methods of removal of high abundant proteins. Here we describe coupling of denaturing ultrafiltration, an alternative depletion strategy, with reverse-phase fractionation and mass spectrometry for characterization of low-molecular-weight proteins and peptides.

Key words LMW proteome, Peptidomics, Biomarker, Ultrafiltration, MALDI, Mass spectrometry

1 Introduction

Mass spectrometric characterization of native peptides and low-molecular-weight (LMW) proteins in biological samples has evolved into a distinct field of peptidomics [1, 2]. Peptidomic studies are attractive because their readout represents complex pathways associated with (patho)physiology of various biological processes [3–5]. The utility of peptidomics ranges from basic biology, through physiology of neuropeptides, to the biomarker discovery in serum and plasma [6, 7]. Even though the interpretation of the screening studies requires caution, the technologies continuously improve and there is considerable interest in the peptidomic applications [8–10]. An efficient extraction of the native peptides is a prerequisite for successful mass spectrometric characterization of the biological systems. The use of denaturing ultrafiltration in peptidomic studies is the focus of this discussion.

It has been estimated that more than 10,000 proteins are present in human plasma or serum [11]. Albumin, the most abundant protein, contributes around 55 % of the total proteome. Twenty-two

Helena Bäckvall and Janne Lehtiö (eds.), *The Low Molecular Weight Proteome: Methods and Protocols*,
Methods in Molecular Biology, vol. 1023, DOI 10.1007/978-1-4614-7209-4_2, © Springer Science+Business Media New York 2013

proteins constitute approximately 99 % of the serum protein content and consequently limit the identification, characterization, and quantification of the lower abundant species [12]. Removal of high abundant proteins narrows the dynamic range and reduces signal suppression in mass spectrometric analysis. A significant challenge for proteomics is to enrich efficiently the lower abundant analytes [13, 14]. A variety of methods to deplete high abundant proteins or to capture specific subsets of peptides/proteins have been developed [15–18]. We focus on the enrichment of the LMW proteins and peptides by ultrafiltration, a fast, easy-to-carry-out, and inexpensive method [19–21]. By using centrifugal membranes with defined molecular weight cutoff (10–50 kDa), proteins with higher molecular weight are retained on the filter while low-molecular-weight species pass through the membrane. The filtrate can be desalted and further fractionated for mass spectrometric characterization. In this chapter, we describe a denaturing ultrafiltration method for enrichment of the native peptides in serum or plasma [7, 22]. The enriched peptides are analyzed by reverse-phase liquid chromatography coupled to MALDI-MS/MS characterization of the analytes. This type of analysis results in our hands in an identification of approximately 250 native peptides from 50 different proteins (*see* **Note 1**).

2 Materials

2.1 Blood Sample Collection and Serum Preparation

1. Vacutainer 10 mL red serum tubes (Becton Dickinson, BD).
2. Vacutainer Brand Safety-Lok Blood Collection Set (BD).
3. Vacutainer Holder (BD).
4. NUNC Cryotubes (BD).

2.2 Ultrafiltration

1. Amicon Ultra centrifugal filter device (15 mL) with 30 kDa MWCO (Millipore) (*see* **Note 2**).
2. Water CHROMASOLV Plus, HPLC grade (Sigma-Aldrich).
3. Guanidine hydrochloride 8 M (*see* **Note 3**).
4. Serum samples (*see* Subheading 2.1).

2.3 Sep-Pak C18 Desalting

1. Sep-Pak C18 cartridge (Waters).
2. Trifluoroacetic acid (TFA) (Sigma-Aldrich) (highly corrosive).
3. Acetonitrile CHROMASOLV Plus, for HPLC ≥99.9 % (Sigma-Aldrich) (highly flammable and harmful).
4. 15 mL centrifuge tubes.
5. Tips.

2.4 Reverse-Phase Liquid Chromatography Fractionation

1. Monolithic C18 4.6 mm × 100 mm (Merck).
2. Agilent 1100 HPLC (*see* **Note 4**).
3. Fraction collector.
4. Tubes.
5. Speed Vac.

2.5 MALDI Mass Spectrometric Peptide Identification

1. a-Cyano-4-hydroxy-cinnamic acid (CHCA), 3.3 mg/mL (Bruker Daltonics) (*see* **Note 5**).
2. MALDI Plate (Applied Biosystems, Inc.).
3. Peptide Calibration Standard (Applied Biosystems, Inc.) (store at −20 °C).
4. Mascot search engine.
5. Mascot Daemon and Mascot Distiller.

3 Method

3.1 Blood Samples Collection and Serum Preparation

1. Blood samples collected in a BD Vacutainer "red-top" tubes are allowed to clot for 60 min (*see* **Note 6**).
2. Serum is separated by centrifugation at 1,200 × *g* for 10 min and aliquoted as soon as possible in convenient volumes (freezing at −80 °C in NUNC Cryotubes if not use immediately) (*see* **Note** 7).
3. Thaw a serum aliquot by submersion in room temperature H_2O; process the sample as soon as possible after thawing (*see* **Note 8**).
4. Spin thawed serum using a bench top centrifuge at ~10,000 × *g* for 10–15 s immediately after thawing to spin down particulates suspended in serum; process without delay on prepared ultrafiltration membranes.

3.2 Ultrafiltration

1. Pipette 2 mL of H_2O into each 30 kDa filter (*see* **Note 9**).
2. Wash filter 5 min at 3,000 × *g* and 10 °C in a refrigerated centrifuge.
3. Pipette 5 mL of 8 M guanidine hydrochloride solution to the washed filter.
4. Add 1 mL serum to the filter and mix by pipetting up and down.
5. Spin for 60 min at 3,000 × *g* and 4 °C in a centrifuge. Stop when the liquid left in the filter is approximately 500 μL (*see* **Note 10**).
6. Collect the filtrate at the bottom of the centrifuge tube for cleanup.

3.3 Sep-Pak C18 Desalting

1. Add 0.6 mL of 0.1 % TFA/acetonitrile to the cartridge spin for 1 min at 100 × *g* and repeat the step twice.
2. Add 1 mL of 0.1 % TFA/water to the cartridge spin it for 2 min at 100 × *g*, and repeat the step for three more times.
3. Apply sample onto the cartridge, spin it for 5 min at 100 × *g*, and reapply the sample once.
4. Wash the cartridge with 1 mL of 0.1 % TFA/water, spin it for 2 min at 100 × *g*, and repeat the step for three more times.
5. Elute peptides with 0.2 mL of 0.1 % TFA/acetonitrile, spin for 1 min at 100 × *g*, and repeat twice (total volume 0.6 mL).
6. Dry eluent in a Speed Vac.

3.4 Reverse-Phase HPLC Separation and Fraction Collection

1. Prepare solvents.
2. Set up the Agilent 1100 HPLC system with a fraction collector. Equilibrate the column with 5 % solvent B at flow rate of 1 mL/min.
3. Resuspend dried filtrate in 30 μL 0.1 % TFA/water. Inject approximately 15 μL of the serum filtrate.
4. Keep the flow isocratic (5 % B) for 2 min, and then start a linear gradient increasing the percentage of solvent B to 40 % over 23 min (*see* **Note 11**). Collect fractions from 2 to 27 min at every 30 s (*see* **Note 12**).
5. Dry fractions for further MS analysis.

3.5 MALDI TOF/TOF Mass Spectrometric Analysis

1. Resuspend each fraction in 1 μL matrix (*see* **Note 13**).
2. Spot 1 μL sample on a MALDI target, together with peptide calibration standards, and allow samples to dry (*see* **Note 14**).
3. Calibrate the 4800 MALDI TOF/TOF mass spectrometer (Applied Biosystems, Inc.).
4. Analyze the samples by acquiring 1,000 shots/spot, laser power 3,000, mass range 800–4,200 Da.
5. Acquire MS/MS spectra of 15 most intense ions in each sample at 1,500 shots/ion, laser power 3,300, 2 kV collision energy with CID gas on.
6. Perform database searches for peptide identification. Processing of the raw MS/MS spectra in Mascot Distiller followed by a Mascot search against the NCBInr human database (no enzyme was specified, MS peptide tolerance 100 ppm and MS/MS tolerance 0.3 Da) allows selection of peptides identified with greater than 99 % confidence (*see* **Note 15**).

4 Notes

1. This chapter describes the preparation of peptides from serum; any biological sample can be processed by denaturing ultrafiltration with appropriate adjustments.
2. The 30 kDa cutoff filters provide efficient elimination of larger proteins and good recovery of peptides. Note that the cutoff is defined by retained species and does not specify what passes the membrane. Typically, peptides with mass lower than the cutoff are eliminated. Recovery of any individual peptide needs to be verified.
3. Denaturation of samples in guanidine hydrochloride gives best recovery of the peptides in our experience.
4. Other HPLC systems and columns are of course appropriate for this reverse-phase fractionation.
5. Dissolve CHCA at 5 mg/mL in 50 % CH_3CN, aliquot 1 mg/vial (0.2 mL), vacuum dry, and desiccate at −20 °C. For experiments, dissolve fresh in 50 % CH_3CN (0.15 mL of dH_2O and 0.15 mL of CH_3CN) at a final concentration of 3.3 mg/mL. Protect matrix from exposure to daylight at all times.
6. Control the time of clotting as closely as possible. Differences in clotting time may affect results. Follow SOP for hazardous biological materials.
7. Be careful not to disturb the interface between serum and red blood cell clot when aspirating serum.
8. (a) It is advisable to keep track of the time from collection to centrifugation and of any lag between centrifugation and freezing. (b) Avoid multiple freeze–thaw cycles. Freeze-thaw cycles of serum samples can cause degradation of peptides and, consequently, changes in the mass spectra.
9. Be careful not to touch the membrane with the tip!!
10. Spin as slow as possible; higher speeds may compromise the separation.
11. Solvent A: 0.1 % TFA in water; solvent B: 0.1 % TFA in acetonitrile. Degas and filter solvents before use.
12. In this experiment, we used a monolithic C18 column which provides good separation in a relatively short time. The gradient can be extended to get better separation depending on the sample.
13. Pipette up and down at least 1 min and swirl around the bottom to dissolve the sample well.

14. Wear gloves to handle the plate. Wait until the spot shrinks to approximately 25 % of its original size (about 10 min) but don't dry the spot completely.
15. Mascot Distiller is an interface to convert mass spectrometry raw data files to MGF files to be used for protein database search through Mascot. Other software compatible with the instrument can be used for the same purpose.

Acknowledgement

This work was supported in part by an Associate Membership from NCI's Early Detection Research Network and R01CA115625 awarded to RG.

References

1. Schulz-Knappe P, Zucht HD, Heine G, Jurgens M, Hess R, Schrader M (2001) Peptidomics: the comprehensive analysis of peptides in complex biological mixtures. Comb Chem High Throughput Screen 4:207–217
2. Falth M, Skold K, Norrman M, Svensson M, Fenyo D, Andren PE (2006) Swepep – a database designed for endogenous peptides and mass spectrometry. Mol Cell Proteomics 5:998–1005
3. Brockmann A, Annangudi SP, Richmond TA, Ament SA, Xie F, Southey BR (2009) Quantitative peptidomics reveal brain peptide signatures of behavior. Proc Natl Acad Sci USA 106:2383–2388
4. Adkins JN, Mottaz H, Metz TO, Ansong C, Manes NP, Smith RD (2010) Performing comparative peptidomics analyses of Salmonella from different growth conditions. Methods Mol Biol 615:13–27
5. Metzger J, Schanstra JP, Mischak H (2009) Capillary electrophoresis-mass spectrometry in urinary proteome analysis: current applications and future developments. Anal Bioanal Chem 393:1431–1442
6. Villanueva J, Nazarian A, Lawlor K, Tempst P (2009) Monitoring peptidase activities in complex proteomes by MALDI-TOF mass spectrometry. Nat Protoc 4:1167–1183
7. Tammen H, Schulte I, Hess R, Menzel C, Kellmann M, Mohring T, Schulz-Knappe P (2005) Peptidomic analysis of human blood specimens: comparison between plasma specimens and serum by differential peptide display. Proteomics 5:3414–3422
8. Davis MT, Auger PL, Patterson SD (2010) Cancer biomarker discovery via low molecular weight serum profiling – are we following circular paths? Clin Chem 56:244–247
9. Ling XB, Sigdel TK, Lau K, Ying L, Lau I, Schilling J, Sarwal MM (2010) Integrative urinary peptidomics in renal transplantation identifies biomarkers for acute rejection. J Am Soc Nephrol 21:646–653
10. Villanueva J, Nazarian A, Lawlor K, Yi SS, Robbins RJ, Tempst P (2008) A sequence-specific exopeptidase activity test (SSEAT) for "functional" biomarker discovery. Mol Cell Proteomics 7:509–518
11. Adkins JN, Varnum SM, Auberry KJ, Moore RJ, Angell NH, Smith RD et al (2002) Toward a human blood serum proteome: analysis by multidimensional separation coupled with mass spectrometry. Mol Cell Proteomics 1:947–955
12. Anderson NL, Anderson NG (2002) The human plasma proteome: history, character, and diagnostic prospects. Mol Cell Proteomics 1:845–867
13. Zolotarjova N, Martosella J, Nicol G, Bailey J, Boyes BE, Barrett WC (2005) Differences among techniques for high-abundant protein depletion. Proteomics 5:3304–3313
14. Whiteaker JR, Zhang H, Eng JK, Fang R, Piening BD, Feng LC et al (2007) Head-to-head comparison of serum fractionation techniques. J Proteome Res 6:828–836
15. Zolotarjova N, Mrozinski P, Chen H, Martosella J (2008) Combination of affinity depletion of abundant proteins and reversed-phase fractionation in proteomic analysis of human plasma/serum. J Chromatogr A 1189:332–338
16. Anderson NL, Jackson A, Smith D, Hardie D, Borchers C, Pearson TW (2009) SISCAPA

peptide enrichment on magnetic beads using an in-line bead trap device. Mol Cell Proteomics 8:995–1005

17. Terracciano R, Pasqua L, Casadonte F, Frasca S, Preiano M, Falcone D et al (2009) Derivatized mesoporous silica beads for MALDI-TOF MS profiling of human plasma and urine. Bioconjug Chem 20:913–923
18. Nilsson J, Ruetschi U, Halim A, Hesse C, Carlsohn E, Brinkmalm G, Larson G (2009) Enrichment of glycopeptides for glycan structure and attachment site identification. Nat Methods 6:809–811
19. Tirumalai RS, Chan KC, Prieto DA, Issaq HJ, Conrads TP, Veenstra TD (2003) Characterization of the low molecular weight human serum proteome. Mol Cell Proteomics 2(10):1096–1103
20. Harper RG, Workman SR, Schuetzner S, Timperman AT, Sutton JN (2004) Low-molecular-weight human serum proteome using ultrafiltration, isoelectric focusing, and mass spectrometry. Electrophoresis 25:1299–1306
21. Zheng X, Baker H, Hancock WS (2006) Analysis of the low molecular weight serum peptidome using ultrafiltration and a hybrid ion trap-Fourier transform mass spectrometer. J Chromatogr A 1120:173–184
22. Orvisky E, Drake SK, Martin BM, Abdel-Hamid M, Ressom HW, Varghese RS et al (2006) Enrichment of low molecular weight fraction of serum for mass spectrometric analysis of peptides associated with hepatocellular carcinoma. Proteomics 6:2895–2902

Chapter 3

Stable Isotope Labeling Methods in Protein Profiling

Johan Lengqvist and AnnSofi Sandberg

Abstract

Mass spectrometry (MS) analysis of peptides and proteins has evolved dramatically over the last 20 years. Improvement of MS instrumentation, computational data analysis, and the availability of complete sequence databases for many species have made large-scale proteomics analyses possible. The measurement of global protein abundance by quantitative mass spectrometry has the potential to increase both speed and impact of biological and clinical research. However, to be able to detect and identify potential biomarkers, reproducible and accurate quantification is essential.

The following chapter describes how to perform quantitative protein profiling using stable isotope labeling methods. Throughout, there is a focus on guidance in selection of an appropriate labeling strategy. With that in mind, we have included a section on acquisition and understanding of the liquid chromatography-mass spectrometry (LC-MS) data format.

Further, we describe the different stable isotope labeling methods and their pros and cons. We start by giving an overview of the overall quantitative proteomics workflow in which extracting relevant biological information from the acquired data is the ultimate goal.

Key words Mass spectrometry, Quantification, Stable isotope labeling

1 Introduction

Liquid chromatography-mass spectrometry (LC-MS) protein profiling ("proteomics") is a quickly maturing scientific discipline. The most advanced labs can now detect over 10,000 proteins in a human sample [1–3]. For biological interpretation of the results, protein identity and protein abundance are of equal importance. Sensitive and accurate protein quantification is important as also small protein-level changes may have biological significance. Further, the altered protein levels are to be put into a context and related protein-level alterations linked for the biological interpretation.

Stable isotope labels are commonly used in mass spectrometry-based quantitative proteomics. This chapter is intended as a primer for understanding the methods and the possibilities of stable isotope labeling for protein profiling.

Helena Bäckvall and Janne Lehtiö (eds.), *The Low Molecular Weight Proteome: Methods and Protocols*, Methods in Molecular Biology, vol. 1023, DOI 10.1007/978-1-4614-7209-4_3, © Springer Science+Business Media New York 2013

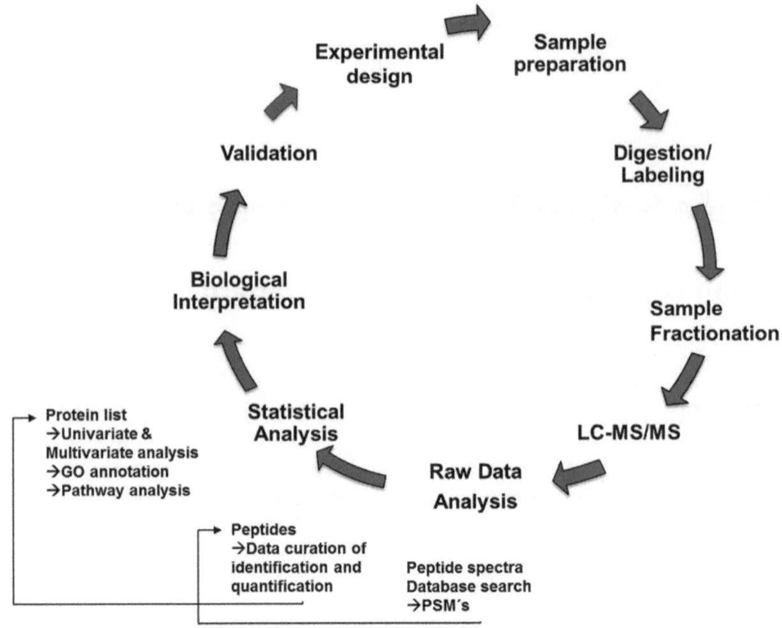

Fig. 1 Schematic overview of a quantitative proteomics workflow. The main sample preparation and data analysis steps are shown

In the first section, we will give a brief description of a quantitative proteomics workflow. Then different methods will be introduced and discussed. Finally a section is dedicated to understanding the LC-MS data format.

1.1 Overview of the Quantitative Proteomics Workflow

An overview of a quantitative proteomics workflow is shown in Fig. 1. The isotopic labels used for protein and peptide quantification are introduced at different steps of the sample preparation depending on the selected label. Metabolic labels are introduced in the cell media, while chemical or enzymatic labeling is done after protein extraction on protein or peptide level. For samples (labeled proteins or peptides) exhibiting a wide dynamic range, fractionation of the sample prior to LC-MS is often a good strategy to reduce sample complexity and thereby increase overall proteome coverage.

After mass spectrometric analysis by LC-MS or LC-MS/MS, the acquired peptide spectra are matched against a database containing theoretical spectra and corresponding peptide sequences by search engine software (for example Mascot [4]). The search output is a list of peptide hits with associated scores as a measure of peptide-spectrum-match (PSM) confidence.

Protein abundance is then inferred from quantitative measurements of the matched peptides. Thus the data analysis process has both a qualitative part (peptide identification) and a quantitative part (protein quantification). Quality control of protein identifications is usually performed by estimating the false discovery rate (FDR) or expected number of false hits as peptide FDR or protein FDR [5], as searches performed with large datasets are associated with false-positive peptide identifications. At the quantitative step in the data analysis, reporter ions/isotopic pairs are compared between samples. Then the peptide ratios are computed and assembled into protein ratios. Assigning peptides to proteins is not a trivial task, as peptide sequences can be shared among several proteins, but crucial for accurate quantification. A tool utilizing the quantitative pattern to match peptides to proteins based on the assumption that peptides originating from the same protein show similar quantitative pattern has been developed and is freely available [6]. The method, which uses correlation analysis, identifies and excludes outliers and detects different protein species, thus improving the information output. This allows correct interpretation of protein forms that are present in the sample but not represented in the database used for protein identification (e.g., unknown splice variants).

Acquired protein ratios are statistically assessed to evaluate the significance of the detected fold changes. In the case of group comparisons and for identifying class discriminating proteins (i.e., a protein or set of proteins that differ between two types of clinical conditions), both univariate and supervised multivariate methods (such as orthogonal partial least squares analysis, OPLS) may be used. To extract meaningful information from these analysis, from which the output often are lists of proteins up- or downregulated in different conditions, further data processing and literature search are usually needed. For extracting biological information from quantitative data from sets of proteins, network building and pathway analysis as well as gene ontology (GO) enrichment analysis may be helpful. Selected proteins may then be validated on a larger material by orthogonal methods.

2 Stable Isotope Labeling Methods

The purpose of the stable isotope label is to serve as an internal standard. An internal standard is used to correct for experimental variation between samples (in other words to normalize the raw data) and is therefore preferably added as early as possible during the sample processing. A stable isotope-labeled peptide is an excellent surrogate standard as it shares all physicochemical properties of the analyte but can be separately detected in the mass spectrometer.

The degree of "multiplexing" in a labeling experiment describes how many samples can be differentially labeled, pooled, and analyzed in one experiment. For instance, in a 4-plex iTRAQ experiment, four different samples can be labeled and pooled

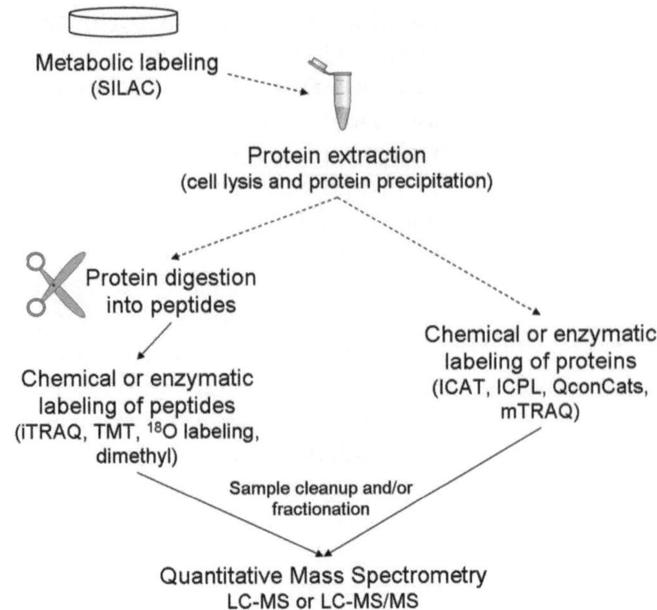

Fig. 2 Schematic showing the stable isotope labels and where they are introduced in the sample preparation workflow, as described in the main text

before final sample processing and analysis. Processing samples together reduces both experimental variation and the LC-MS instrument time required for data acquisition. It also increases coherence in the resulting data.

An overview of the stable isotope labels described in the text and their respective point of application is shown in Fig. 2. Metabolic labels are supplemented in the cell media, and thus introduced prior to cell lysis and protein extraction. After disruption of cells and cell debris removal, protein concentration is determined. Often an additional cleanup of the sample such as protein precipitation in acetone is required prior to labeling. In this case the protein determination step is done after the precipitation and solubilization in appropriate buffer. A defined amount is then submitted either to labeling using isotopic labels designed for labeling of intact proteins or to protein digestion into peptides followed by labeling of the peptides. The enzymatic cleavage step might be followed by an additional peptide determination and/or a 1D-SDS-PAGE to verify complete trypsinization. The labeled sample is then submitted to a cleanup step where excessive labeling reagents, salts, and/or detergents that might interfere with subsequent MS analysis are removed. After this, the sample is freeze-dried and dissolved in LC-MS-compatible buffer.

Stable isotope tags are often used in discovery studies aiming to detect differences in protein amounts under given conditions. In such experiments, two samples, one labeled with a "light", and

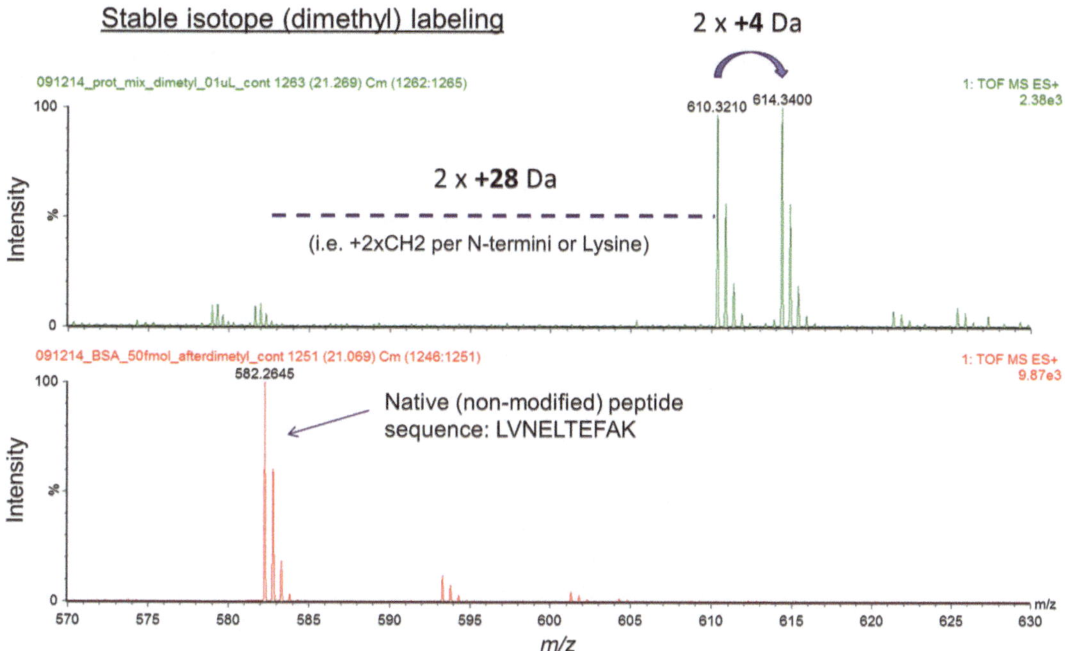

Fig. 3 An example of a stable isotope labeling experiment using the dimethyl-labeling approach. Equal amounts of a tryptic digest were labeled and pooled. Shown are the spectra for the labeled sample (*upper panel*) and the non-labeled peptide (*lower panel*). Per N-terminal amine and for each lysine residue the label adds two methyl groups (CH_2 or CD_2), i.e., +28 Da (light) or +32 Da (heavy form). The peptide is the LVNELTEFAK peptide from a tryptic digest of bovine serum albumin

one with a "heavy" isotope tag, are mixed and analyzed by mass spectrometry. The relative abundance of each sample can be determined by comparing their relative ion intensities. This is because peptides from differentially labeled samples share all other properties but will differ in mass. This is illustrated in Fig. 3. The upper panel shows a mass spectrum where equal amounts of a peptide have been labeled with a "light" and a "heavy" label (+4 Da) and pooled at a ratio of 1:1. As seen the intensity of the two peaks are highly similar. The lower panel shows the mass spectrum of the non-labeled peptide. The tag adds mass to the peptide (in this case +2 × 28 Da corresponding to the addition of two methyl groups, i.e., +2 × CH_2). The heavy version of the tag instead adds +2 × 30 Da (by adding 2 × CD_2 instead of CH_2). For this type of labeling method, as the difference is observed in the mass spectrum without fragmenting any peptide, the quantification is done in single-stage mass spectrometry mode, or MS1.

In contrast, for the isobaric stable isotope labeling reagents (e.g., the "isobaric tag for relative and absolute quantification," iTRAQ, and "tandem mass tags," TMT, see below), all forms of the tag have the same mass. So no mass shifts are observed in the MS1 spectrum. Instead each label gives a different fragment, a "reporter" ion, when the peptide is fragmented in the mass spectrometer (i.e., in tandem mass spectrometry mode, or "MS2"). Such reporter ion fragments are often observed in the low mass range of the MS2 spectrum (i.e., in the 110–150 m/z range). A detailed discussion of the different modes of data acquisition is included below.

Both relative and absolute quantitative measurements can be made using stable isotope labels. In protein profiling experiments designed for high coverage (e.g., in biomarker discovery experiments), relative quantification is often used. Here quantification of peptides/proteins is based on comparing isotopically "light" and "heavy" forms of the whole sample. The relative ratios of the peak intensities give the relative abundance of the samples.

For absolute quantification, analyte abundance is obtained through relating analyte response to that of a surrogate standard (e.g., a synthetic peptide) of known concentration. For each analysis a response curve is generated to verify a quantitative assay. Absolute quantification is often employed as a targeted analysis, i.e., focusing on a single or a few proteins only. A targeted approach is for instance more likely in a validation phase of a study (*see* Fig. 1) where a few (biomarker) candidates have to be analyzed in a large number of samples.

Below different stable isotope labeling methods are discussed and exemplified. Choosing the most appropriate labeling method is a balance between factors such as cost, LC-MS instrument time available, as well as the sample throughput needed. Further, a specific type of mass spectrometer may be more suited for one particular isotope labeling method. Finally the aim of the study, e.g., in terms of depth of analysis needed, is important in choosing an experimental design.

2.1 Protein Labeling

Introducing the isotope label early in the sample processing is beneficial to control for as much experimental variation as possible. Protein labels have the advantage of labeling before tryptic digestion. Variable enzymatic digestion is often neglected in proteomics and remains poorly explored. So it is attractive to be able to control for this step in sample processing.

2.1.1 Metabolic Labeling

Stable Isotope Labeling by Amino Acids in Cell Culture

Cell culture enrichment of proteins with stable isotope-labeled amino acids (SILAC) is in many ways the preferred labeling method as it is introduced at a very early stage and can control for all steps of sample processing and analysis [7]. Labeling is done by supplementing cell growth medium with one or several isotope-labeled (in other words heavy) amino acids. As the medium has been depleted of the corresponding natural amino acid, cellular proteins will be enriched for the labeled amino acid after a number of cell doublings (generally 4–5 doublings). For example, if heavy arginine is used, all normal arginine residues (having ^{12}C atoms) will have been replaced by stable isotope-labeled arginines (that have instead for instance six ^{13}C atoms). Each arginine in the SILAC-labeled sample will then be 6 Da heavier than for a non-labeled sample. Depending on the number of arginines (n) in its sequence, each protein will be $n \times 6$ Da heavier. After mixing labeled and non-labeled samples, the relative amount of heavy and light protein can be determined. In proteomics experiments, trypsin is often used to digest proteins into peptide. As trypsin cleaves after arginine and lysine residues in the protein sequence, a high percentage of all tryptic peptides will include arginine. So in the example above where stable isotope-labeled arginine was used, such a peptide will be 6 Da heavier than the same tryptic peptide from a non-labeled sample. Several heavy amino acid labels besides arginine have been investigated for SILAC including lysine, tyrosine, methionine, and leucine [8–11]. Lysine-labeling is equally suited as arginine for use with trypsin digestion.

The isotope composition of the labeled amino acid(s) used will determine the mass increase of a labeled peptide. For instance, stable isotope-labeled arginine can be bought that is +4, +6, +10, +11, or +17 Da heavier than normal arginine. By combining for instance the +10 and +17 Da variants with a non-labeled sample, a triplex experiment can be designed, where three different samples are pooled and analyzed as one. In theory any combination of SILAC-labels can be used. The mass separation must be large enough that differentially labeled peptides do not overlap in the mass spectrum. In practice, +4 Da is often the minimum used. When selecting the labeled amino acid reagent, one should also consider which isotope to use. Some isotopes alter the retention time of the peptides more than others. For instance, the retention time shift compared to a non-labeled peptide is more pronounced with ^{2}H (deuterium, D) than for ^{13}C or ^{15}N.

When applying SILAC it is critical that labeling is complete. Incomplete labeling will give rise to subpopulations and thus leads to erroneous quantification. Further, some cell lines are capable of metabolic conversion of modified amino acids (for instance conversion of isotopic arginine to proline) [12]. Accordingly, labeling efficiency has to be verified by LC-MS (MS1) analysis.

In terms of data acquisition, quantification of SILAC-labeled samples is done based on the ion intensity in the mass spectrum (MS1). SILAC increases the complexity of the sample. Instead of one peptide form, two or three are seen in the mass spectrum, comparable with the two peaks in the example shown in Fig. 1a. Due to this, SILAC benefits from the use of high-resolution instruments such as Orbitraps or FTICR instruments.

SILAC was developed [10] and still is mostly used for labeling cells grown in culture. However, several higher organisms have also been labeled including the generation of a SILAC mouse [13] and fruit fly [14]. For a review of the applications of metabolic labeling to various organisms *see* ref. 15. However, many organisms, most notably humans, are unsuitable to maintain an isotope-enriched (SILAC) diet. One of the drawbacks of SILAC has been that it cannot be applied to clinical studies. In an effort to overcome this, the "super-SILAC" mix method has been published [16], in which a mix of different cell lines is used to make an isotope-labeled standard.

2.1.2 ICAT

One of the first stable isotope labeling methods for quantitative proteomics was the isotope-coded affinity tags (ICAT) [17]. It consists of labeling cysteine amino acid residues with a thiol-specific isotope-coded (light and heavy version) biotin-tag. Protein samples are mixed and after proteolysis, tagged peptides can be affinity purified. After purification and optional sample fractionation, the ion intensity ratio for heavy- and light-labeled peptide pairs is obtained from the LC-MS analysis. The affinity purification step greatly reduces sample complexity. For the 344,855 theoretical tryptic peptides from the 6,113 yeast proteins, only 30,619 (<10 %) contain a cysteine residue [17]. As the sample is made less complex, coverage (depth) of analysis should increase. Further, the cysteine-enrichment adds information for the protein identification process as all identified peptides should contain a cysteine.

A drawback of the ICAT-method is that not all potentially interesting peptides will be covered. Important protein features (e.g., specific posttranslational modifications in non-cysteine peptides) may be excluded from analysis.

2.1.3 Isotope-Coded Protein Label

The isotope-coded protein label (ICPL) labeling method targets all amine groups (lysine residues and protein N-termini) at the protein level. It was introduced in 2005 [18] and additional publications are appearing [19, 20]. The method now allows 4-plex quantification experiments. For a review of the method *see* ref. 21. One advantage is that samples are labeled and then pooled early in the sample processing. Thus protein separation steps can be included to improve the depth of analysis. Fractionation at the protein level is very efficient for reducing complexity of the peptide mixture since it occurs before enzymatic digestion. This is because a single protein can give well over a 100 peptides when digested.

2.1.4 QConCats

QConCats [22, 23] are artificial proteins genetically engineered to include the peptides of choice for each experiment. The plasmid constructs are then expressed (for instance, in bacterial system) to incorporate stable isotope-labeled amino acids. Once generated, QConCAT-expression plasmids are a renewable source of labeled internal standard peptides. In principle, the method is related to the AQUA-peptide approach described below. However the labeled protein can be spiked very early in the sample workup and thus standardize for the tryptic digestion step. One such protein has been engineered specifically to serve as an internal quality control standard, the QCAL protein [24]. The QCAL protein contains peptides that can be used to control for mass spectrometer resolution, mass accuracy, and linearity as well as the RPLC-separation.

2.1.5 Intact Protein iTRAQ Labeling

Although mostly used for peptide labeling, the iTRAQ reagents can be used to label intact proteins [25–27]. Using the 4-plex version of the iTRAQ label, the method allows a high degree of multiplexing among the protein-level labeling methods.

2.2 Peptide (Post-digestion) Labeling

Labeling peptides after the digestion step is more efficient than protein labeling because sites are more accessible for labeling than in intact proteins. However, for these methods digestion efficiency is not controlled for since the samples are labeled and pooled after the enzymatic digestion step. If digestion is not reproducible, this will introduce experimental variability for the protein abundances measured. It is therefore good practice to check the protein digestion step. This is most easily done by running an SDS-PAGE gel of the protein sample before and after digestion (after digestion no protein bands should be visible). Further, the peptide concentration after digestion should be measured and equal amounts labeled.

2.2.1 ^{18}O-Labeling

This method is based on incorporation of ^{18}O-isotopes from heavy ($H_2{}^{18}O$) water at peptide carboxy termini by proteases such as trypsin [28, 29]. Complete incorporation gives two ^{18}O-atoms per peptide and a mass shift of +4 Da. After digestion and labeling, trypsin activity has to be neutralized to avoid back-exchange reactions. This can be done by, e.g., adding acid, heating the sample, removal of protease by molecular weight cutoff filters, or using bead-linked (immobilized) trypsin [30, 31]. Due to the relatively small mass shift, ^{18}O-labeling is better suited for MALDI analysis (giving mostly singly charged peptide ions) than for electrospray ionization (ESI). For multiply charged ions, the risk of overlap between the ^{16}O- and ^{18}O-labeled peptide peaks is higher.

2.2.2 Dimethyl Labeling

The great advantage of the dimethyl-labeling method [32] is low reagent cost (~0.1 Euro for labeling 25 μg of protein in a triplex experiment) [33, 34]. This means that large amounts of starting material can be labeled. A large amount of starting material is highly advantageous when performing several steps of sample preparation and fractionation to compensate for sample losses in each step.

The method is based on reductive dimethylation of the amine groups of peptide N-termini and lysine residues. Dimethyl labeling is done as a duplex or a triplex labeling experiment. As shown in Fig. 3, dimethyl labeling gives +28 or +32 Da per modified residue compared to the native peptide. Straightforward protocols for labeling in solution, during solid phase extraction (SPE) cleanup, or online to the LC-MS analysis have been published [34]. Applications include in-depth phosphoproteome analysis using anti-tyrosine antibody affinity capture [35] as well as an analysis of primary human leukocytes [36]. Most recently, reductive dimethylation was combined with selective peptide removal (the so-called TAILS-method) to study protease cleavage in complex biological samples including bronchoalveolar lavage fluid [37].

As for many stable isotope labeling methods (ICAT, SILAC, ICPL), quantification in dimethyl labeling experiments is performed in MS1. Several freely available software packages can perform MS1 quantification. These include the MSQuant software [38], msInspect that is part of the Trans-Proteomic Pipeline (TPP) package [39], and the MaxQuant software package [40]. Both MSQuant and msInspect have the advantage compared to MaxQuant that they can be used for all MS data, not only high-resolution data from FTICRs and Orbitraps.

2.2.3 Isobaric Labels: Tandem Mass Tags (TMT) and Isobaric Tags for Relative and Absolute Quantification (iTRAQ)

Isobaric labels consist of three parts: a *linker* (contains a reactive group; TMT and iTRAQ have an *N*-hydroxysuccinimide ester group that makes them reactive towards N-terminal amines and the amine group of lysine residues), a *reporter* (with a mass unique for each tag), and a *balance* group (adjusted so that the net mass is identical among the tags). Peptides labeled with isobaric tags are thus equal in mass and will be observed as a single peak in the mass spectrum (MS1). Upon fragmentation, the reporter group of the tag is released and the corresponding reporter ions are observed in the low mass region (m/z range 113–121 for iTRAQ, and 126–131 for TMT). Quantification is then based on the relative intensities of those reporter ions. Thus, to be quantified, a peptide has to be fragmented to generate MS/MS (MS2) data. One advantage is that several samples are pooled and the amount of peptide "available" will be greater when samples are combined. Also the sample will not be more complex than a non-labeled sample.

Quantification in MS2 (as compared to quantification in MS1) using isobaric labels has the advantage that both peptide identification and quantification can be performed on the same tandem mass spectrum, simplifying optimization. Another benefit is higher signal-to-noise ratios. However, efficient peptide fragmentation and good resolution in the selection of precursor ions for fragmentation is important, as the presence of unidentified fragments in the MS2 spectra reduces ion score which has a negative effect on the identification rate. In addition, reporter ions from co-selected precursor ions superimpose on the "true" reporter ions, skewing the quantification ratios towards 1 [41]. The negative effect on protein identification rates due to the occurrence of unidentified fragments has been shown to be more pronounced using 8-plex iTRAQ compared to 6-plex TMT and 4-plex iTRAQ [42]. This is contradicted by a study showing higher quantification accuracy using 8-plex iTRAQ compared to 4-plex iTRAQ without reduced protein identification rates [43]. A new software-based approach to remove interfering signals by identifying the reporter ions via the accurate mass differences within a single tandem mass spectrum has been presented [44]. This will filter out near-isobaric signals, but will obviously not have any effect on superimposed reporter ions in the quantification analysis.

The frequency of co-selected precursor ions increases with sample complexity, and is consequently reduced by sample pre-fractionation [45]. MS instrument developments that increase scan rate, resolution and improve fragmentation have been implemented [46–48]. These improve the quantification of complex samples, but co-selection of precursor still occurs.

Both TMT and iTRAQ have a high degree of multiplexing compared to other isotopic labels which makes them popular. For iTRAQ, the original 4-plex has been extended to eightfold multiplexing [49], while TMT allows sixfold multiplexing [50–52]. By using an internal reference standard, multiplexing beyond the number of available tags is made possible.

Additional considerations that one should be aware of are side reactions. In the original publication, the authors stated low degrees (<3 %) of tyrosine derivatization and of un-reacted N-terminal and lysine amines [53]. Further, there is the possibility of isotope contamination (isotopic impurities). If these are present the reporter ion ratios will be skewed and thus the entire quantification [54].

3 Targeted Protein Quantification: MRM/SRM Analysis

To validate proposed biomarker candidates, targeted analytical approaches capable of high throughput are needed. For targeted assays, triple quadrupole (QQQ) instruments are often used. Such instruments have excellent sensitivity when set to record only a limited number of specific ions. If the QQQ is set to record only selected fragment masses for a given peptide mass, the analysis becomes very specific. Such an analysis will be very sensitive and able to tolerate very complex samples where the risk of overlapping peaks is high. This mode of analysis is called selected reaction monitoring (SRM) or multiple reaction monitoring (MRM). Here the MRM term will be used throughout.

The internal standards used for targeted protein quantification are often synthetic stable isotope-labeled peptides. These are commercially available as AQUA-peptides (for "absolute quantification"). One drawback of the AQUA-peptide approach is high cost and the fact that not all peptides may be easy or even possible to synthesize.

MRM methodology has proven very robust with interlab CVs <15 % [55]. Importantly, the method shows good correlation with ELISA-based methods [56, 57]. As described below the sensitivity of MRM-assays is constantly increasing. All these factors make MRM-based approaches an attractive alternative to ELISA and Western blot methods.

The true test of sensitivity and assay specificity is always plasma analysis due to the very wide range of plasma protein abundances [58]. Without abundant protein depletion and without peptide fractionation, the level of quantification (LOQ) is ~1–20 μg of protein per mL of plasma [59]. Removing the six most abundant plasma proteins (~85 % of the protein mass) by immunodepletion increases sensitivity by about sevenfold to 0.14–2.5 μg/mL [59]. Sensitivity can be increased further to 1–10 ng/mL by minimal peptide fractionation [60]. Antibody-based peptide enrichment, i.e., the SISCAPA-approach, can further improve sensitivity [61]. Such methods can now reach pg/mL level proteins in plasma with minimal sample processing from 1 mL of starting material. Small sample volumes (~10–50 μl of plasma) have a detection limit in the ng/mL range [62, 63]. While holding great promise, a major obstacle is the generation of the anti-peptide affinity reagent. However, recently an automated screening of monoclonal antibodies for SISCAPA-assays has been published [64].

In a complex sample (e.g., in the absence of affinity purification or fractionation) one limiting factor of an MRM-assay is background signal interference. For low-level analysis, assays (i.e., the transitions monitored) should be validated, by testing multiple transitions per analyte peptide. Transition is the word used for each specific combination of intact peptide mass and the associated specific fragment measured. In other words, "monitoring the transition 614 > 118" means fragmenting the mass 614 and recording the signal for the *m/z* 118 fragment (compare Fig. 9). One novel instrument improvement, MRM^3-capability (pronounced "MRM-cubed"), yields increased sensitivity over "standard" MRM (LOQ ~10 ng/mL in non-depleted human plasma) [57]. This analysis effectively covers an abundance range of ~6-fold compared to the most abundant protein in plasma. The added specificity (and thus improved depth of analysis) is due to the ability to further fragment ions in an MS^3-experiment (MS/MS/MS) which further reduces the chance of signal interference.

One bottleneck of targeted proteomics approach is the assay development phase. MRM/SRM-assay development involves three steps. (1) Finding the best (i.e., most LC-MS "friendly") peptides for each protein. Ideally each protein should be quantified based on more than one peptide. (2) For the peptide(s) selected, the best analytical (mass spectrometry) parameters and the most intense fragment ions have to be found. Finally, (3) the developed MRM/SRM assay has to be tested using a real-life sample to establish sensitivity (specificity) against the sample background. A useful review of these steps is found in ref. 65. See also the data acquisition section below.

Once found the specific parameters in terms of peptide chosen and fragment ions measured can be utilized by anyone. Thus an MRM/SRM assay for a given peptide can be adopted by everyone with a triple quadrupole who wants to measure that peptide. Accordingly, public repositories of validated MRM-assays are now being generated, primarily for yeast [66] but also for mouse and human. Recently a large-scale effort to generate validated MRM-assays for yeast was published [67]. In this chapter the assay generation and testing was automated based on extensive (low-cost) peptide synthesis. The throughput of the optimization process was reportedly >100 peptides/h which in theory enables generation of MRM-assays on a proteome-wide scale. Multiplexed MRM-analysis strives to measure as many peptides as possible in a single LC-MS run. To achieve this the analysis can be time-scheduled such that each peptide is only measured at the precise time it elutes from the LC-column.

Triple quadrupoles are relatively cheap instruments. This together with excellent dynamic range, ease of use, and high throughput makes them attractive for peptide quantification in proteomics experiments. Triple quadrupole use is sure to increase in the proteomics field.

4 Acquisition and Understanding of the Liquid Chromatography-Mass Spectrometry (LC-MS) Data Format

4.1 Understanding Data Acquisition

For proteomic analysis the final separation step is very often reversed-phase liquid chromatography and most instruments are sold as integrated LC-MS systems. Such a system has three basic parts: (1) the liquid chromatograph, (2) the LC-MS coupling, and (3) the mass spectrometer itself. Analytical performance is depending on all three parts working together effectively.

4.1.1 Basic Parts of a Mass Spectrometer

A mass spectrometer has three basic parts. Those are the ion source, one or several mass analyzers, and the ion detector. Mass measurement is based on the separation of charged species (ions) in time and/or in space. In the instrument (under vacuum) ions can be manipulated, e.g., accelerated, focused, directed, and stored using electric lenses, high magnetic fields, and other means. This is possible because in vacuum the ions will not collide with gas molecules as they would in air.

In the ion source ions are generated from neutral molecules or already formed ions are brought into the gas-phase. For peptide and protein analysis two ionization methods, electrospray ionization (ESI) and matrix-assisted laser desorption ionization (MALDI), dominate. Both greatly facilitate the analysis of fragile organic molecules such as DNA/RNA, proteins/peptides, and endogenous metabolites. ESI is a process of solvent evaporation and droplet fission ending with gas-phase ions. It is influenced by many factors including flow rate, source temperature, and solvent volatility. Because lower flow rate gives higher sensitivity for electrospray [68], LC-MS in proteomics is often done at very low flow rates (<1,000 nL/min for the so-called nanoLC-MS). The mass analyzer separates ions of different mass. Several different types exist but describing each is beyond the scope of this chapter. An instrument that has two or more mass analyzers is called a tandem mass spectrometer (or an MS/MS instrument). If the mass analyzers are of different types the instrument, such as the QTOF, is called a hybrid instrument (in that case a combination of a quadrupole analyzer and a time-of-flight analyzer).

4.2 Understanding LC-MS Data: Quantification in MS1

The mass spectrum is an abundance plot of mass-separated ions (Fig. 4a). The y-axis shows ion counts (intensity) and is often labeled in percentage relative intensity (compared to the most abundant peak in the spectrum). Looking to the top right of the spectrum, we see the absolute intensity value (in this case $1.07e^4$ for the m/z 582 ion). The mass scale (the x-axis) shows mass-to-charge ratio (m/z), i.e., atomic mass units divided by the charge of the ion. Figure 4a shows an electrospray mass spectrum of a peptide. The peak at m/z 582.2645 corresponds to the doubly charged peptide LVNELTEFAK from a tryptic digest of bovine serum albumin. The actual mass of the peptide is $[582.2645 \times 2] - 2 = 1{,}162.529$. Several tools exist for calculating the peptide mass from the amino acids sequence (e.g., the MS-Product tool available at http://prospector.ucsf.edu/prospector). In acidic solutions amine groups will be protonated and positively charged. Because trypsin cleaves after arginine and lysine residues, tryptic peptides will be able to carry at least two positive charges, i.e., both at the peptide N-terminus and the arginine or lysine side chain.

An LC-MS experiment also has a time dimension. In LC-MS, mass spectra are recorded continuously as the LC-separation is running. Each time point in the LC chromatogram corresponds to a single mass spectrum recorded at that time. When a peptide elutes from the column and is recorded it will be observed as a peak in the LC-chromatogram. Figure 4b shows a nanoLC-MS analysis of a BSA digest with the LC peak corresponding to the LVNELTEFAK peptide indicated. In addition to the LVNELTE FAK peptide peak many peptides from the BSA digest are present. They all come from the same protein so they should in theory be present at the same concentration. However, peak intensities are

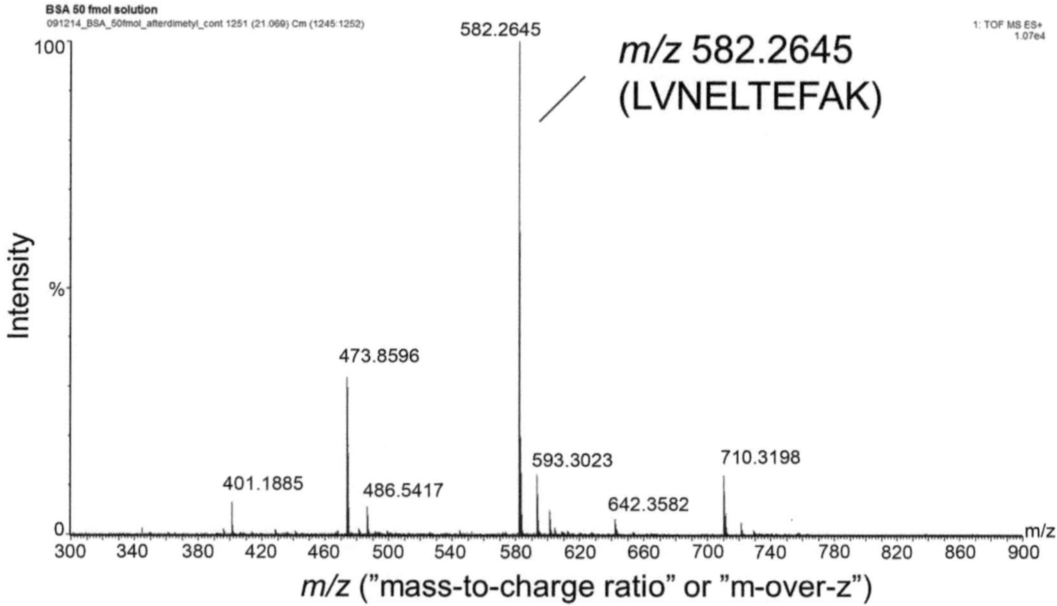

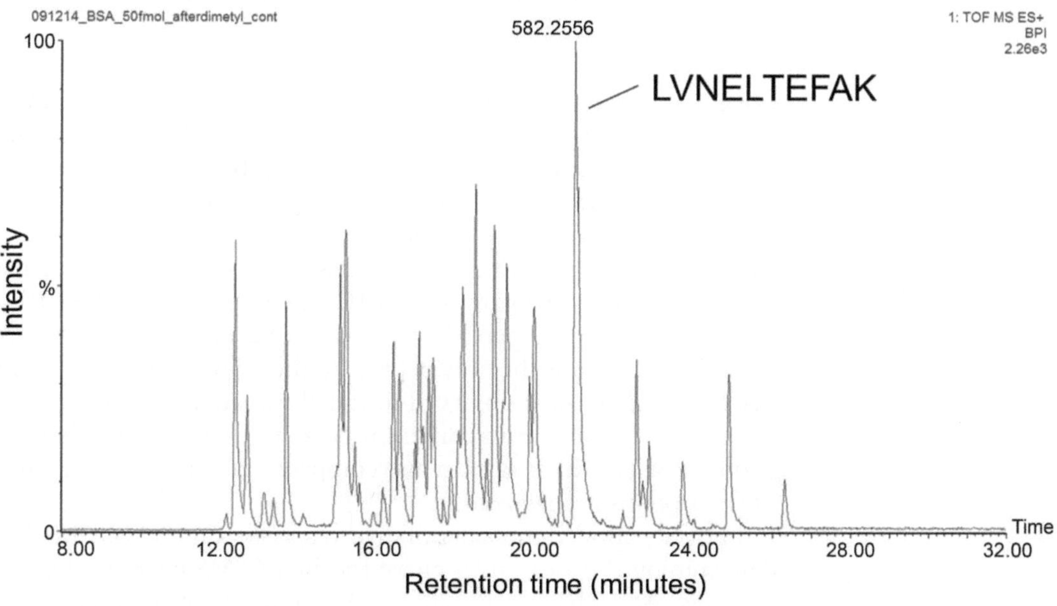

Fig. 4 (**a**) The mass spectrum of the LVNELTEFAK peptide from a tryptic digest of bovine serum albumin (BSA). The doubly charged intact peptide is observed at a mass-to-charge ratio (*m/z*) of 582.2645. A few peptides of lower intensity are also observed. (**b**) An LC-MS chromatogram of a tryptic digest of bovine serum albumin. The LC-peak of the LVNELTEFAK peptide is indicated. (**c**) Peptide quantification in LC-MS (MS1). The *upper panel* shows the extracted ion chromatogram of the LVNELTEFAK peptide (*m/z* 582.265). The mass window used was 100 mDa (0.1 Da). Below is shown the integration of the ion chromatogram giving an area under curve of 963

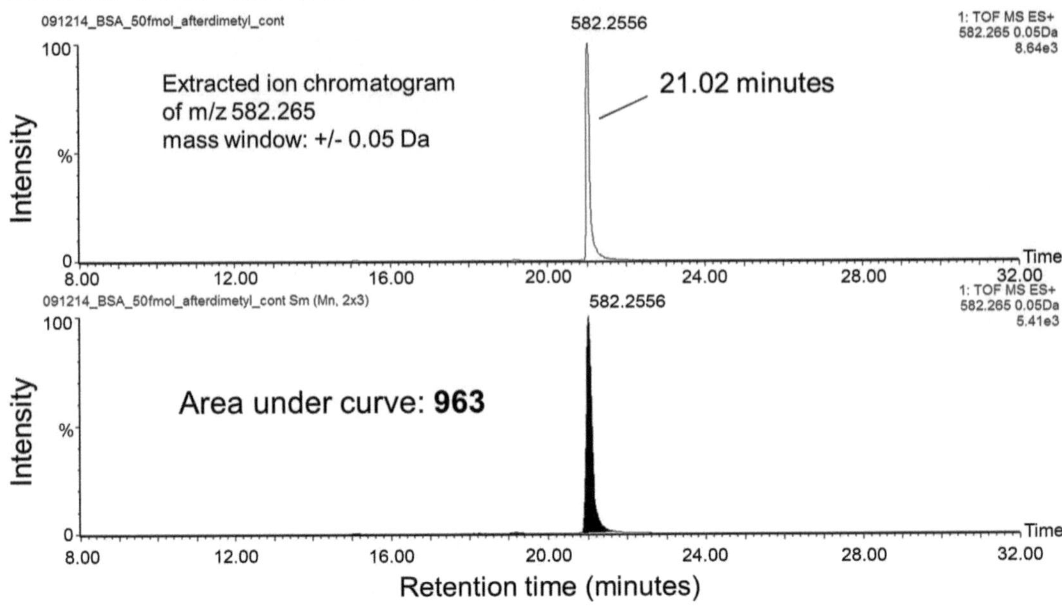

Fig. 4 (continued)

not equal. This is because signal intensity depends on the chemical composition (i.e., on the sequence) of each peptide. It follows that signal intensity cannot be directly compared between two different peptides. This is why the best internal standard is an isotopically labeled form of the same peptide.

If only the signal for a specific mass value is plotted this is called an extracted ion chromatogram (EIC). Figure 4c shows the EIC of the LVNELTEFAK peptide ion (*m/z* 582.2645 ± 0.05 Da). As LVENELTEFAK is the only peptide in the BSA digest that has this mass, a single peak is observed in the chromatogram. However, if the sample contains many different proteins there is a risk that two peptides will have the same mass (compare Fig. 5a, c where a digest of six proteins is analyzed).

For a peptide, the quantitative response can be obtained by determining the area under curve for the LC-MS peak (in this case 963, Fig. 4c). This is also called to integrate the peak. As a rule of thumb about 15 data points (individual spectra) should be recorded across the peak for sufficient quantitative accuracy. Thus it may be necessary to adjust the scan speed (the time used to record a single mass spectrum) depending on the speed of the chromatographic separation. In Fig. 4c we see that the LVNELTEFAK peak is about 20 s wide. As the scan speed was 1 s, in this case 15–20 measurements

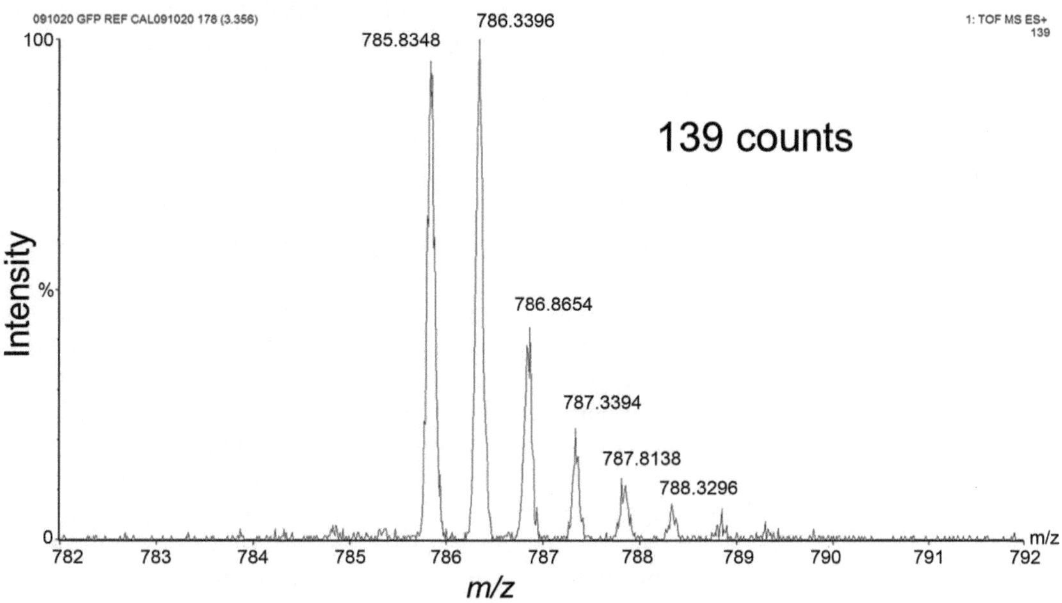

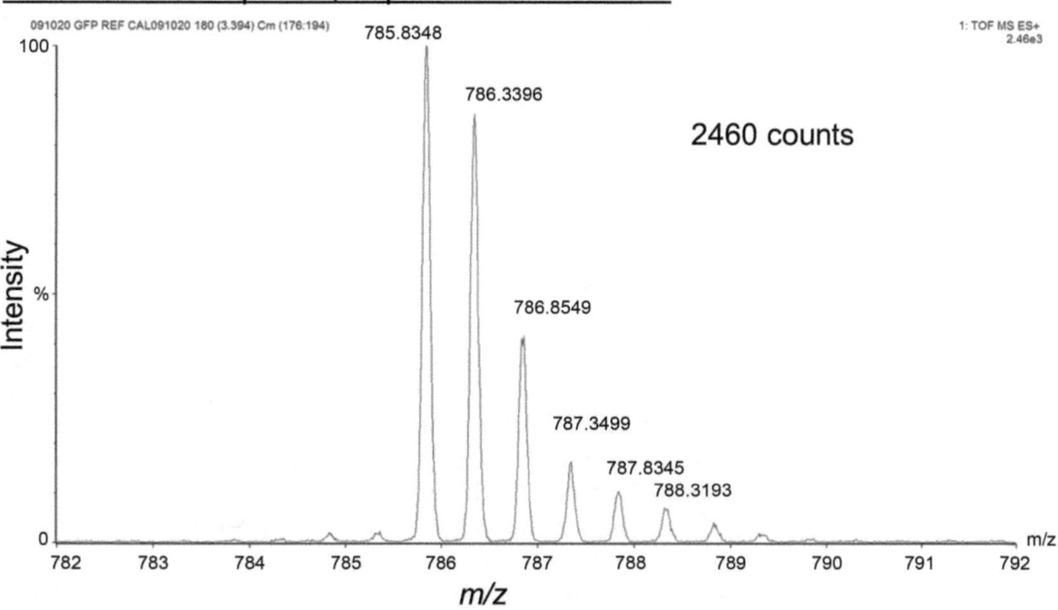

Fig. 5 Improved ion statistics will give improved peak shapes and thus improved mass accuracy and quantitative accuracy. In (**a**) is shown a single mass spectrum (scan) of the Glu-Fibrinogen peptide standard where the most intense ion has 139 counts. In (**b**) is shown the combination of 20 consecutive spectra which gives 2,460 counts

will define the peak. Making several measurements will improve quantitative accuracy as well as mass accuracy.

The area under the peak can be determined either for the peak in the *mass spectrum* (as in Fig. 5) or in the *LC-MS chromatogram* as shown in Fig. 4c. In the iTRAQ and TMT-methods described below, the reporter ion peaks in the MS/MS spectrum are integrated similar to Fig. 5. Almost all other methods, including targeted protein measurements by MRM and in MS1-based quantification, quantify based on integration of the peak in the LC-MS chromatogram as in Fig. 4. High mass resolution minimizes the risk of overlapping peaks. A non-resolved overlapping peak will distort the peak shape of the analyte and thus shift the observed mass value. Quantitative values will also be skewed if overlapping peaks are measured together.

The quantitative value (or peak intensity) can be measured in several ways. One way is simply to report the peak top height (i.e., the highest recorded ion intensity). This is sometimes useful for very weak signals. Another way is to measure the area under the peak. This is called to integrate the peak (compare Fig. 4c).

If a peak is of low intensity, the peak outline is rugged (Fig. 5a shows a single spectrum of the Glu-Fibrinogen peptide). If the data is noisy, data-processing such as smoothing of the peak outline may be needed. If more ions are recorded, the peak shape improves drastically and the peak outline becomes less noisy (compare Fig. 5b, 20 summed scans).

For quantification very low-intensity peaks can be difficult to integrate. Also very high-intensity peaks can be problematic because at some point the detector will reach saturation. This means that even if the analyte concentration continues to increase signal intensity will not accurately reflect the concentration increase. The response will plateau out (compare Fig. 6c). Thus it is important to determine the (dynamic range) limitations of the system you are using. This is done by recording a standard curve using one or several analytes of known concentration and checking over what range the standard curve is linear (*see* Fig. 6a, b). Figure 6 shows a dilution curve for a standard peptide where the same samples were analyzed on a QTOF (a) or a QQQ instrument (b). Because the QTOF instrument was run in nanoLC-MS mode, the whole dilution curve (12 points) took 17 h to complete. The same 12 samples were done in 48 min on the triple quadrupole (using a much higher flow rate, 400 μL/min) illustrating the higher throughput possible when using high chromatography flow rates.

However, the nanoLC-MS analysis is more sensitive. This is illustrated by the fact that the dilution curve in A extends below 1 fmol peptide.

Further, triple quadrupoles have better linearity of response and a greater dynamic range. Comparing Fig. 6a, b, the trend lines indicate a more linear response for the QQQ-instrument (Fig. 6b). Further, the QTOF analysis shows detector saturation at higher signal intensities. Figure 6c shows the ion chromatograms for 2 and 250 fmol peptide loaded (indicated by the arrows in Fig. 6a). At 250 fmol loaded, the peak is flat on top (due to detector saturation)

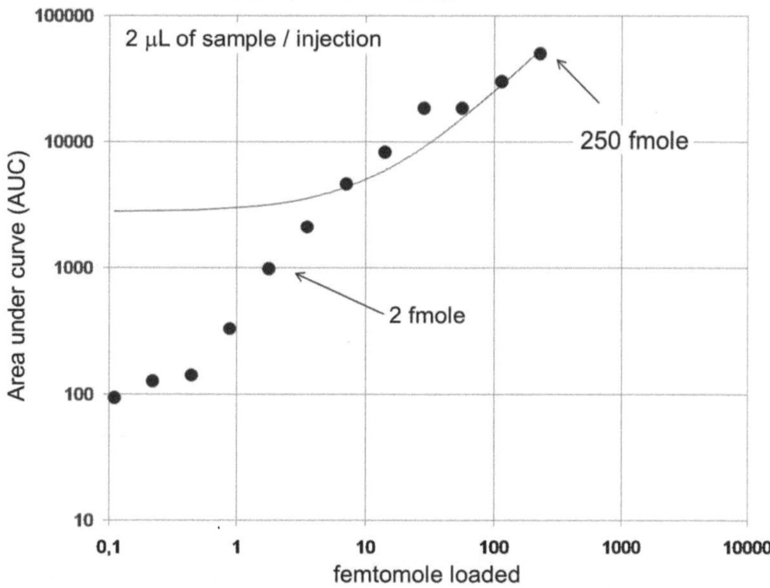

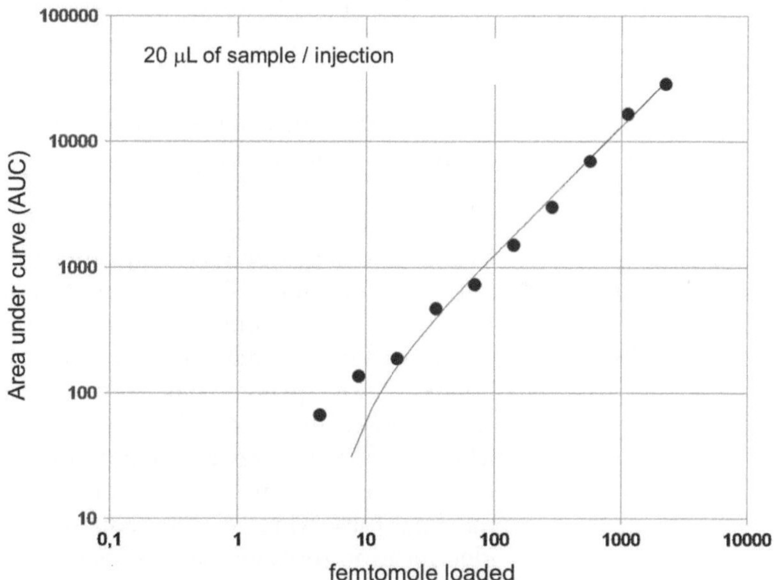

Fig. 6 Peptide concentration response curves generated on two different instruments (nanoLC-QTOF and a high flow rate triple quadrupole in (**a**) and (**b**), respectively). *Trend lines* indicate the fit to a theoretical linear response. In (**c**) is shown the extracted ion chromatograms from the nanoLC-QTOF analysis for 2 and 250 fmol points. At 250 fmol, detector saturation is evident

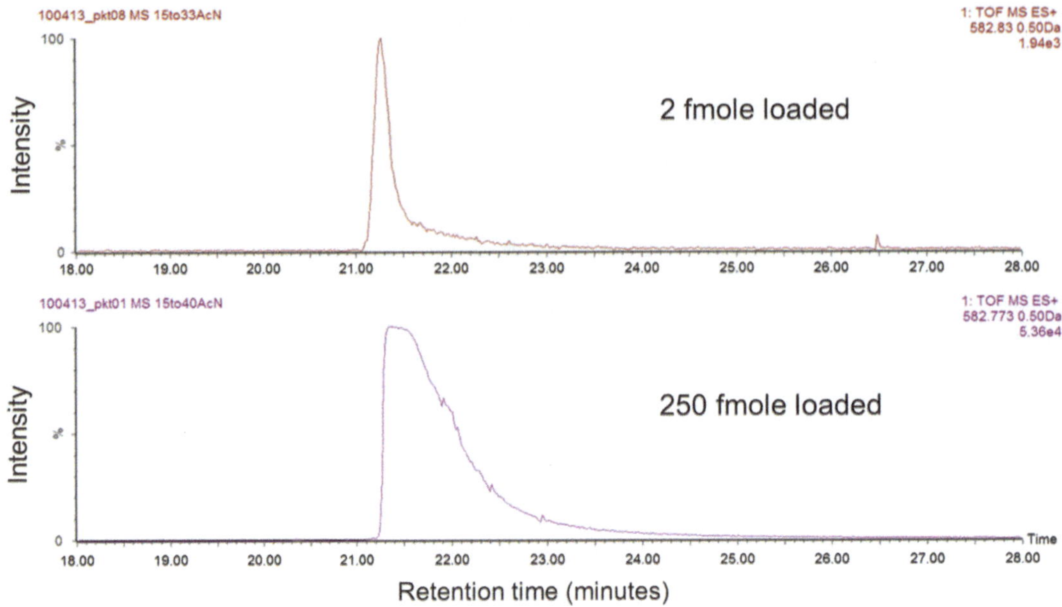

Fig. 6 (continued)

and not Gaussian. The response will not accurately reflect increased concentration in this range.

4.3 Understanding MS/MS Data: Quantification in MS2

The peptide bond is relatively weak. So in MS/MS of peptides, fragment ions will be seen that differ by the mass of single amino acid residues. Figure 7a shows the fragment ion spectrum for the LVNELTEFAK peptide with the peptide sequence ions indicated. The sequence ions have different intensities because in a peptide, bonds are not equally strong, some will break more easily than others. Importantly, a complete set of fragment ions covering the whole peptide sequence may not always be seen. For instance, in Fig. 7a, the fragment corresponding to the final "AK" dipeptide is missing. Observing all the sequence ions will aid in sequence identification through database searching and of course in manual spectrum interpretation.

For the database search the mass accuracy of the recorded data has to be specified, both for the precursor (in MS1) and for the fragments (in MS2). In Fig. 7 the mass errors for some of the fragment ions are given. As the errors are all <0.012 Da, a cutoff of for example ±0.02 Da would be appropriate for a database search with this particular spectrum.

It is important to know the mass accuracy of the data. In data evaluation it is good practice to plot the mass error values for the whole population of identified peptides. This will at a glance give

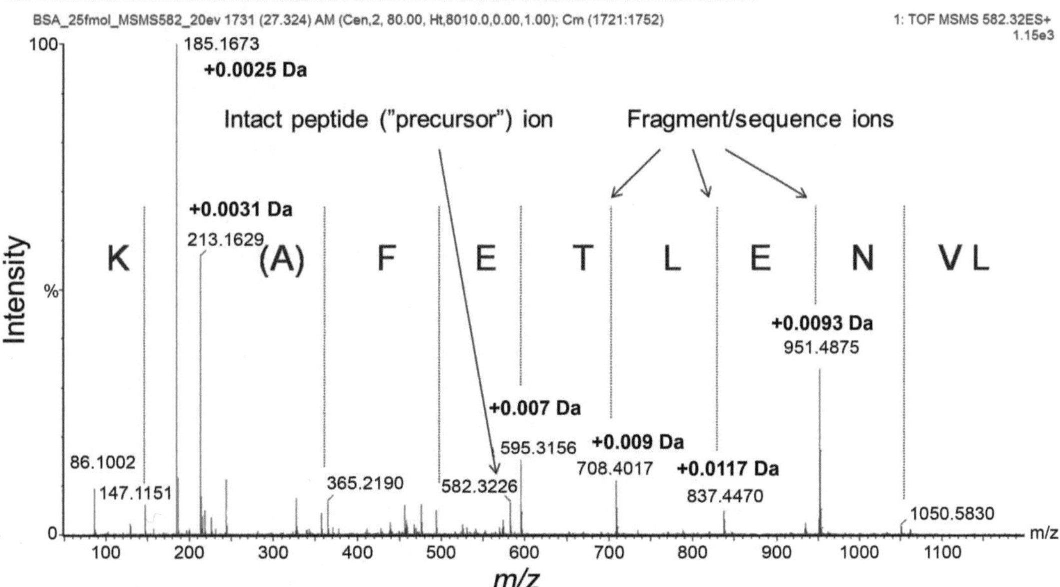

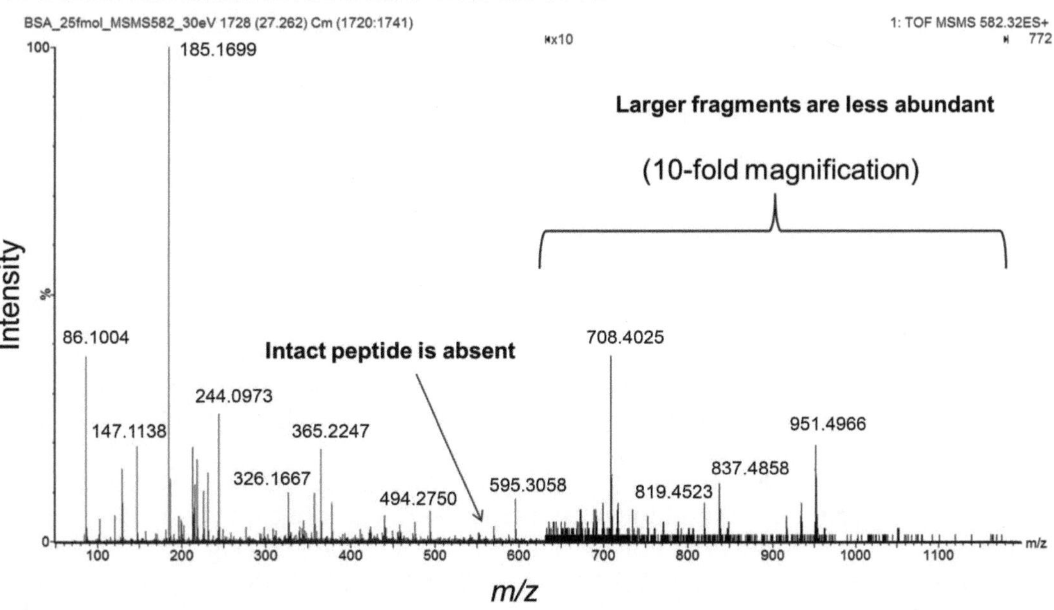

Fig. 7 Fragmentation (MS/MS) spectrum of the LVNELTEFAK peptide from bovine serum albumin at 20 (**a**) and 30 eV of fragmentation energy (**b**). In (**a**) the peptide sequence is indicated and the numbers (in *italics*) give the mass error relative to the theoretical mass of each fragment. The intact peptide (*m/z* 582) is indicated by the *arrow*. In (**b**) the high mass region is shown at tenfold magnification

an understanding of the mass accuracy for the data set. Several database search tools including X!Tandem (http://www.thegpm.org/tandem [69]) and Phenyx (http://www.genebio.com/products/phenyx [70]) automatically provide such plots.

Experimental conditions will influence how the MS/MS spectrum looks. Under low-energy conditions (20 eV) of fragmentation energy (Fig. 7a), a small peak is still observed for the intact peptide at *m/z* 582. With higher fragmentation energy (30 eV), the intact peptide is not seen and the high mass fragments are less intense indicating more extensive fragmentation (Fig. 7b). For peptide identification purposes, having fragment ions covering the whole length of the peptide sequence is ideal.

So peptide fragmentation patterns (the relative intensity of the MS/MS ions) will vary depending on the experimental conditions. Further, different instruments may have other modes of fragmentation besides the collision-induced dissociation (CID) method of the QTOF shown above. Fragmentation conditions should be checked and optimized for each instrument and ideally for each peptide when using stable isotope labeling method that are quantified from MS/MS spectra. Such methods include iTRAQ, AQUA, and TMT-tags. Importantly, traditional ion trap instruments have limited sensitivity in the low mass range due to instrument design and are not suited for iTRAQ- or TMT-based studies. LTQ-Orbitrap hybrid instruments incorporate an ion trap mass analyzer but novel fragmentation methods (high-energy C-trap dissociation, electron transfer dissociation) allow the use of isobaric tags [46, 47, 71–75].

AQUA-peptides for targeted quantification are based on MS/MS data but unlike iTRAQ quantification is not done on reporter ions. Instead each peptide is quantified based on the respective signal of selected fragment (sequence) ions. This makes AQUA-peptide quantification perfectly suited for triple quadrupole instruments that are relatively sensitive compared to TOFs, ion traps, or Orbitraps when measuring only a few selected fragments.

4.4 MRM/SRM Assays: Quantification by Selective Fragmentation in MS2

For best sensitivity in an MRM-analysis, the best instrument parameters and the most intense fragment ions should be found for each peptide. For the LVNELTEFAK example (Fig. 7a) the fragment ions at *m/z* 185.167, 213.163, and 951.488 are good candidates as they are relatively intense. However, when choosing the best fragment, not only intensity matters. A fragment above the mass of the precursor (in this case above *m/z* 582) will generally have less noise. So the 951 fragment may give a better signal-to-noise ratio for complex biological samples.

If the peptide digest is complex, peptides may have the same or very similar mass. Figure 8a shows the extracted ion-chromatogram for the mass (*m/z*) 614.3 from a dimethyl-labeling experiment of a tryptic digest mixture of six standard proteins including BSA

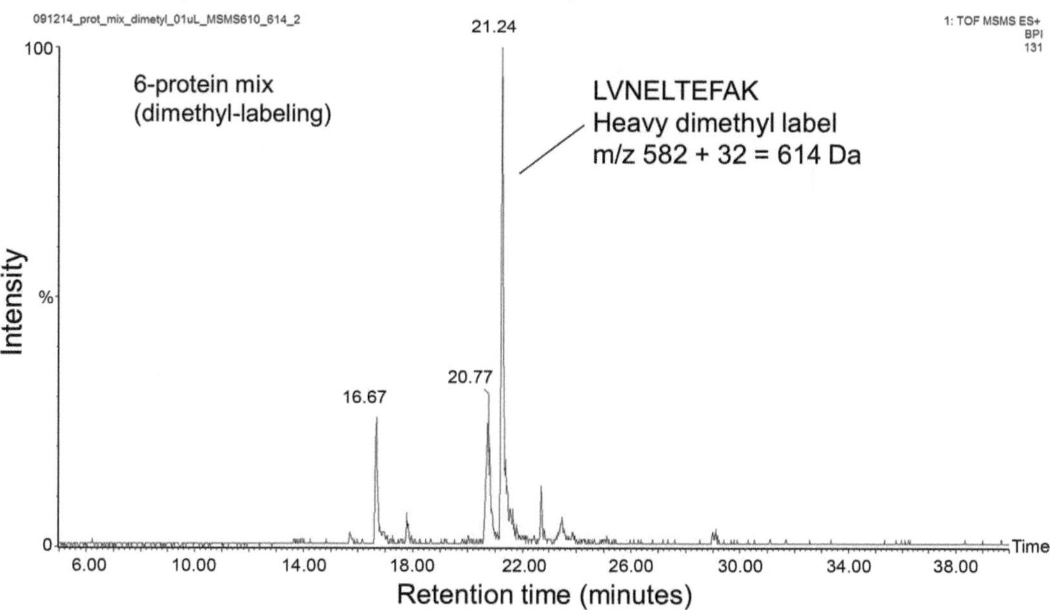

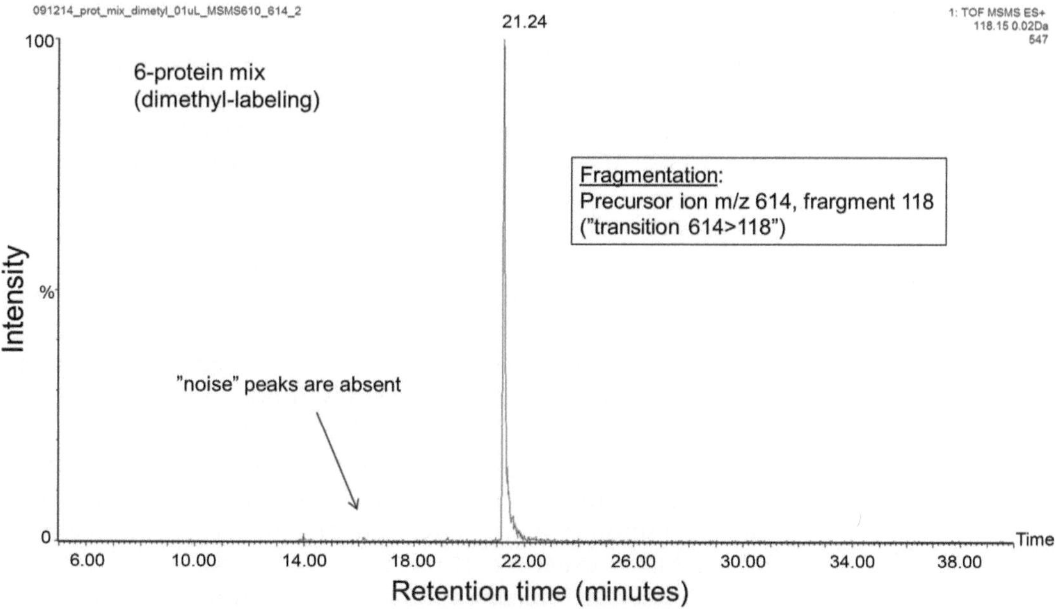

Fig. 8 Increased specificity of analysis by measuring a given peptide–fragment ion combination. In (**a**) is shown the ion chromatogram for the peptide ion (*m/z* 614) from an analysis of a 6 protein digest. The LVNELTEFAK peptide is indicated. In (**b**) is shown the ion trace for the *m/z* 118 fragment of the peptide and in (**c**) the MS/MS spectra of the heavy and light dimethyl-labeled forms (*upper* and *lower panel*, respectively). In each panel the mass of the precursor (intact peptide) is indicated

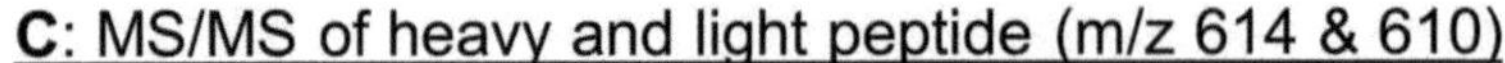

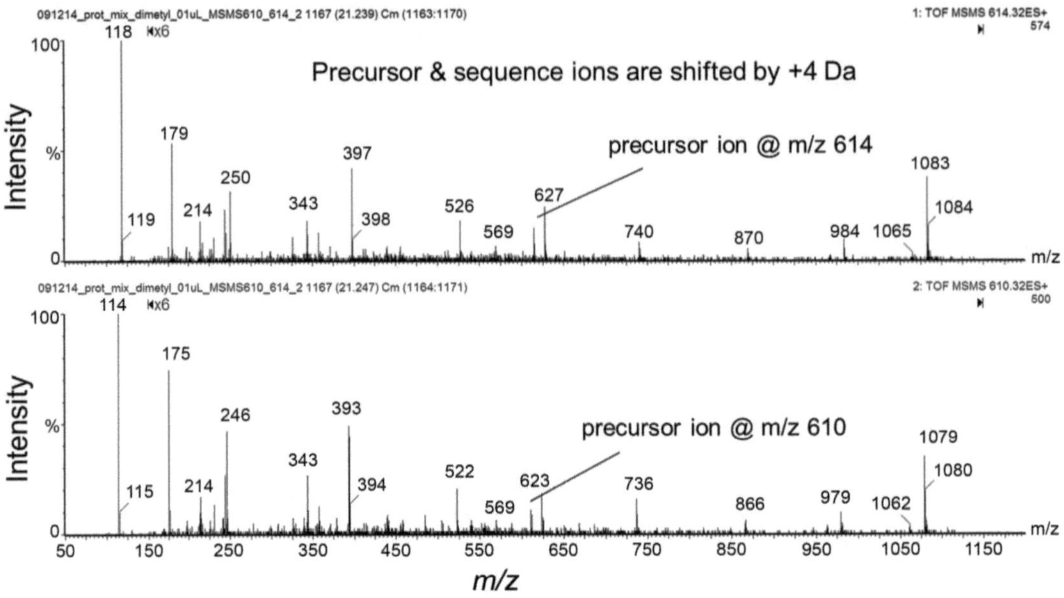

Fig. 8 (continued)

(compare Fig. 3). The labeled LVNELTEFAK peptide elutes at ~21.24 min. Even with only six proteins in the mixture we can see that other peaks of mass 614.3 are also observed, e.g., at retention time 16.67. If we did not know the correct retention time, we would not be able to say which LC-peak corresponds to the correct peptide. However, even if two different peptides have the same mass, the MS/MS fragments may not be identical. For instance, if the intensity (ion trace) for the fragment at *m/z* 118 is plotted, a peak is only observed at 21.24 min (Fig. 8c). Comparing with the EIC in Fig. 8a, the peak at 16.67 min is not seen. So although the peak at 16.67 min has a mass of 614.3, it does not give a fragment at *m/z* 118 in MS/MS. So an MS/MS experiment can increase the specificity of the analysis and make it easier to pick out a specific peptide in a complex sample. As mentioned previously, this type of data acquisition is highly suited for triple quadrupole instruments. For an MRM experiment, the first quadrupole would be set to transmit only the precursor ion (*m/z* 614.3). In the collision cell, all ions that have a mass of 614 are fragmented. The final quadrupole is set to transmit only the correct fragment to the detector (*m/z* 118). In essence, these two consecutive mass filters put a rigorous constraint on what is finally detected. To make the analysis even more specific, the last quadrupole can be switched between two or more fragment masses from the same peptide (hence the name "multiple reaction monitoring"). As seen in the MS/MS spectrum (Fig. 8b), the ions at *m/z* 397, 627 or 1,083 are all good candidates for measuring the heavy dimethyl-labeled form of LVNELTEFAK (Fig. 8c, upper panel).

When peptides are differentially labeled using stable isotope tags, both the mass of the intact peptide and the fragment ions will change. For instance, if the peptide is instead labeled with the light dimethylation reagent, we see that both the precursor (*m/z* 610) and all the fragment (sequence) ions are shifted to lower mass (−4 Da, Fig. 8c. lower panel).

5 Data Quality Control

Because of the extremely complex samples and instrument variation, experimental data for protein profiling will never be perfect when many samples are compared. Thus some form of quality control is good practice as will be outlined below.

Consecutive LC separations may not behave exactly alike. For instance, shifts in retention time and ion intensity can occur. Nanospray stability is particularly hard to maintain over time. Further the masses measured may drift over time. Finally, the sample itself can change over time. Even if samples are maintained in refrigerated autosampler units, peptide losses may occur. These factors combine to cause drifts in retention time, mass, and ion intensity over time.

Many software packages do peak integration and it is important to keep in mind that the algorithms used will vary. It is not certain that the software will manage to match a given peak in one run to the same peak in all other runs. For instance, an LC peak may show tailing to the right. Across many runs, the integration of the tail may vary from sample to sample. Alternatively, a second peak may be present shortly after the first. With chromatography shifts, the second peak can be adequately resolved in some samples while in others the peaks may co-elute. To overcome such problems some form of retention time alignment between runs is often done by the software prior to peak matching across runs. In short, software-based (automated) protein quantification is a complex process involving many steps where errors may occur that will propagate to affect the quantification result. So to have the best confidence, the final results should be manually inspected and verified.

One efficient way to determine variation in the instrumental (LC-MS) system is to use quality control (QC) samples. These are technical replicates of the same sample that are included among the real study samples. NanoLC-MS runs often last an hour or more. This means that for a large study, data may be acquired over several days or even weeks. In such a setting QC samples should be included at regular intervals to monitor instrument drift. When analyzed, runs of the QC sample should show low variation in

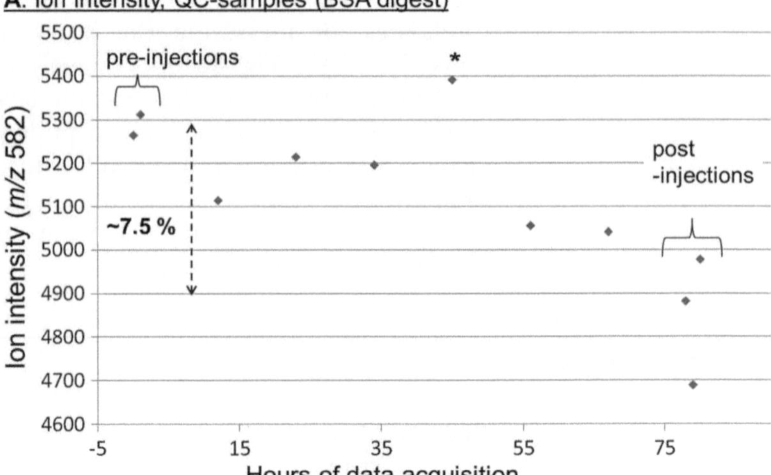

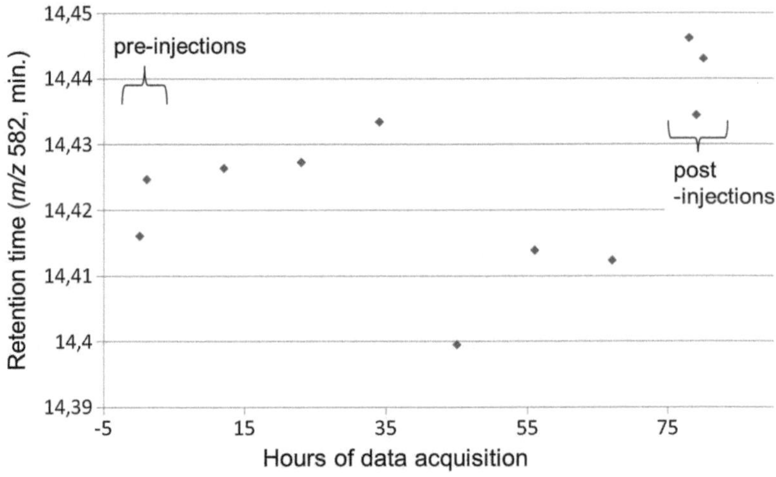

Fig. 9 Using quality control samples (a BSA digest) to detect instrument drift over extended analyses. The ion intensity (**a**) and retention time (**b**) of the LVNELTEFAK peptide (*m/z* 582) from the quality control sample is plotted against time for the duration of data acquisition. The QC sample was injected after every tenth sample analyzed

terms of peak retention time, mass value, and ion intensity. Variation should be calculated, e.g., in the form of standard deviation or coefficient of variance, and the response should be plotted over time from the first injection to the last.

In Fig. 9a the ion intensity of the LVNELTEFAK peptide from a BSA digest sample is plotted over time. The BSA digest was used as a quality control sample in a biomarker discovery study and was injected after every tenth real sample. Data acquisition was over a period of 2.5 days (~72 h) with 12 QC samples injected in total over that time. Both the ion intensity (*m/z* 582) and the retention time of the LVNELTEFAK peptide are varying over the course of the data acquisition (Fig. 9a, b respectively). It is evident that there is covariation between the ion intensity and the retention time shifts (see, e.g., the sixth injection indicated by the asterisk in Fig. 9a, b). However, the QC samples themselves will allow intensity and retention time adjustments if needed. Importantly, without the QC samples we would not be aware of the intensity declining over time.

6 Conclusion

Quantitative protein analysis using LC-MS-based methods is a rapidly evolving field. It is clear that a general consensus of experimental methods has so far not been reached, and that proteome coverage is still far from "global." However, recently published work is starting to set a benchmark in terms of proteome coverage achieved. High proteome coverage, in terms of both protein identities and of protein quantities, is important to extend the analysis beyond the high abundance proteins.

In this chapter, we have given an overview of the most widely used stable isotope labeling methods and their main pros and cons. In addition, we have described the principles of how mass spectrometry-based quantification is performed, illustrated with practical examples. We believe that this knowledge will aid in the selection of suitable quantitative method as well as trouble shooting, and hope that this will lead to quantitative data that will serve as a solid starting point for downstream biological interpretation.

References

1. Aebersold R (2009) A stress test for mass spectrometry-based proteomics. Nat Methods 6:411–412
2. Nilsson T, Mann M, Aebersold R, Yates JR III, Bairoch A, Bergeron JJ (2010) Mass spectrometry in high-throughput proteomics: ready for the big time. Nat Methods 7:681–685
3. Nagaraj N, Kulak NA, Cox J, Neuhauser N, Mayr K, Hoerning O, Vorm O, Mann M (2012) System-wide perturbation analysis with nearly complete coverage of the yeast proteome by single-shot ultra HPLC runs on a bench top Orbitrap. Mol Cell Proteomics 11(M111):013722
4. Perkins DN, Pappin DJ, Creasy DM, Cottrell JS (1999) Probability-based protein identification by searching sequence databases using mass spectrometry data. Electrophoresis 20:3551–3567
5. Reiter L, Claassen M, Schrimpf SP, Jovanovic M, Schmidt A, Buhmann JM, Hengartner MO, Aebersold R (2009) Protein identification false discovery rates for very large proteomics data sets generated by tandem mass spectrometry. Mol Cell Proteomics 8:2405–2417
6. Forshed J, Pernemalm M, Branca RM, Sandberg A, Lehtiö J (2011) Enhanced information and improved accuracy from shotgun proteomics by protein quantification based on peptide quality control (PQPQ). Mol Cell Proteomics 10(10):4
7. Zhang G, Fenyo D, Neubert TA (2009) Evaluation of the variation in sample preparation for comparative proteomics using stable isotope labeling by amino acids in cell culture. J Proteome Res 8:1285–1292
8. Ibarrola N, Molina H, Iwahori A, Pandey A (2004) A novel proteomic approach for specific

identification of tyrosine kinase substrates using [13C]tyrosine. J Biol Chem 279:15805–15813

9. Martinovic S, Veenstra TD, Anderson GA, Pasa-Tolic L, Smith RD (2002) Selective incorporation of isotopically labeled amino acids for identification of intact proteins on a proteome-wide level. J Mass Spectrom 37:99–107

10. Ong SE, Blagoev B, Kratchmarova I, Kristensen DB, Steen H, Pandey A, Mann M (2002) Stable isotope labeling by amino acids in cell culture, SILAC, as a simple and accurate approach to expression proteomics. Mol Cell Proteomics 1:376–386

11. Ong SE, Kratchmarova I, Mann M (2003) Properties of 13C-substituted arginine in stable isotope labeling by amino acids in cell culture (SILAC). J Proteome Res 2:173–181

12. Bendall SC, Hughes C, Stewart MH, Doble B, Bhatia M, Lajoie GA (2008) Prevention of amino acid conversion in SILAC experiments with embryonic stem cells. Mol Cell Proteomics 7:1587–1597

13. Kruger M, Moser M, Ussar S, Thievessen I, Luber CA, Forner F, Schmidt S, Zanivan S, Fassler R, Mann M (2008) SILAC mouse for quantitative proteomics uncovers kindlin-3 as an essential factor for red blood cell function. Cell 134:353–364

14. Sury MD, Chen JX, Selbach M (2010) The SILAC fly allows for accurate protein quantification in vivo. Mol Cell Proteomics 9:2173–2183

15. Gouw JW, Krijgsveld J, Heck AJ (2010) Quantitative proteomics by metabolic labeling of model organisms. Mol Cell Proteomics 9:11–24

16. Geiger T, Cox J, Ostasiewicz P, Wisniewski JR, Mann M (2010) Super-SILAC mix for quantitative proteomics of human tumor tissue. Nat Methods 7:383–385

17. Gygi SP, Rist B, Gerber SA, Turecek F, Gelb MH, Aebersold R (1999) Quantitative analysis of complex protein mixtures using isotope-coded affinity tags. Nat Biotechnol 17:994–999

18. Schmidt A, Kellermann J, Lottspeich F (2005) A novel strategy for quantitative proteomics using isotope-coded protein labels. Proteomics 5:4–15

19. Maccarrone G, Turck CW, Martins-de-Souza D (2010) Shotgun mass spectrometry workflow combining IEF and LC-MALDI-TOF/TOF. Protein J 29:99–102

20. Turtoi A, Mazzucchelli GD, De Pauw E (2010) Isotope coded protein label quantification of serum proteins—comparison with the label-free LC-MS and validation using the MRM approach. Talanta 80:1487–1495

21. Kellermann J (2008) ICPL – isotope-coded protein label. In: Posch A (ed) 2D PAGE: sample preparation and fractionation. Humana, Totowa, NJ, pp 113–123

22. Pratt JM, Simpson DM, Doherty MK, Rivers J, Gaskell SJ, Beynon RJ (2006) Multiplexed absolute quantification for proteomics using concatenated signature peptides encoded by QconCAT genes. Nat Protoc 1:1029–1043

23. Rivers J, Simpson DM, Robertson DH, Gaskell SJ, Beynon RJ (2007) Absolute multiplexed quantitative analysis of protein expression during muscle development using QconCAT. Mol Cell Proteomics 6:1416–1427

24. Eyers CE, Simpson DM, Wong SC, Beynon RJ, Gaskell SJ (2008) QCAL–a novel standard for assessing instrument conditions for proteome analysis. J Am Soc Mass Spectrom 19:1275–1280

25. Wiese S, Reidegeld KA, Meyer HE, Warscheid B (2007) Protein labeling by iTRAQ: a new tool for quantitative mass spectrometry in proteome research. Proteomics 7:340–350

26. Prudova A, auf dem Keller U, Butler GS, Overall CM (2010) Multiplex N-terminome analysis of MMP-2 and MMP-9 substrate degradomes by iTRAQ-TAILS quantitative proteomics. Mol Cell Proteomics 9:894–911

27. auf dem Keller U, Prudova A, Gioia M, Butler GS, Overall CM (2010) A statistics-based platform for quantitative N-terminome analysis and identification of protease cleavage products. Mol Cell Proteomics 9:912–927

28. Mirgorodskaya OA, Kozmin YP, Titov MI, Körner R, Sönksen CP, Roepstorff P (2000) Quantitation of peptides and proteins by matrix-assisted laser desorption/ionization mass spectrometry using ^{18}O-labeled internal standards. Rapid Commun Mass Spectrom 14:1226–1232

29. Fenselau C, Yao X (2009) 18O2-Labeling in quantitative proteomic strategies: a status report. J Proteome Res 8:2140–2143

30. Petritis BO, Qian W-J, Camp DG, Smith RD (2009) A simple procedure for effective quenching of trypsin activity and prevention of 18O-labeling back-exchange. J Proteome Res 8:2157–2163

31. Sevinsky JR, Brown KJ, Cargile BJ, Bundy JL, Stephenson JL Jr (2007) Minimizing back exchange in 18O/16O quantitative proteomics experiments by incorporation of immobilized trypsin into the initial digestion step. Anal Chem 79:2158–2162

32. Ji C, Guo N, Li L (2005) Differential dimethyl labeling of N-termini of peptides after guanidi-

nation for proteome analysis. J Proteome Res 4:2099–2108

33. Boersema PJ, Aye TT, van Veen TA, Heck AJ, Mohammed S (2008) Triplex protein quantification based on stable isotope labeling by peptide dimethylation applied to cell and tissue lysates. Proteomics 8:4624–4632
34. Boersema PJ, Raijmakers R, Lemeer S, Mohammed S, Heck AJ (2009) Multiplex peptide stable isotope dimethyl labeling for quantitative proteomics. Nat Protoc 4:484–494
35. Boersema PJ, Foong LY, Ding VM, Lemeer S, van Breukelen B, Philp R, Boekhorst J, Snel B, den Hertog J, Choo AB, Heck AJ (2010) In-depth qualitative and quantitative profiling of tyrosine phosphorylation using a combination of phosphopeptide immunoaffinity purification and stable isotope dimethyl labeling. Mol Cell Proteomics 9:84–99
36. Raijmakers R, Heck AJ, Mohammed S (2009) Assessing biological variation and protein processing in primary human leukocytes by automated multiplex stable isotope labeling coupled to 2 dimensional peptide separation. Mol Biosyst 5:992–1003
37. Kleifeld O, Doucet A, auf dem Keller U, Prudova A, Schilling O, Kainthan RK, Starr AE, Foster LJ, Kizhakkedathu JN, Overall CM (2010) Isotopic labeling of terminal amines in complex samples identifies protein N-termini and protease cleavage products. Nat Biotechnol 28:281–288
38. Mortensen P, Gouw JW, Olsen JV, Ong S-E, Rigbolt KTG, Bunkenborg J, Cox JR, Foster LJ, Heck AJR, Blagoev B, Andersen JS, Mann M (2009) MSQuant, an open source platform for mass spectrometry-based quantitative proteomics. J Proteome Res 9:393–403
39. Bellew M, Coram M, Fitzgibbon M, Igra M, Randolph T, Wang P, May D, Eng J, Fang R, Lin C, Chen J, Goodlett D, Whiteaker J, Paulovich A, McIntosh M (2006) A suite of algorithms for the comprehensive analysis of complex protein mixtures using high-resolution LC-MS. Bioinformatics 22:1902–1909
40. Cox J, Mann M (2008) MaxQuant enables high peptide identification rates, individualized p.p.b.-range mass accuracies and proteome-wide protein quantification. Nat Biotechnol 26:1367–1372
41. Karp NA, Huber W, Sadowski PG, Charles PD, Hester SV, Lilley KS (2010) Addressing accuracy and precision issues in iTRAQ quantitation. Mol Cell Proteomics 9:1885–1897
42. Pichler P, Kocher T, Holzmann J, Mazanek M, Taus T, Ammerer G, Mechtler K (2010) Peptide labeling with isobaric tags yields higher identification rates using iTRAQ 4-plex compared to TMT 6-plex and iTRAQ 8-plex on LTQ Orbitrap. Anal Chem 82:6549–6558
43. Pottiez G, Wiederin J, Fox HS, Ciborowski P (2012) Comparison of 4-plex to 8-plex iTRAQ quantitative measurements of proteins in human plasma samples. J Proteome Res 11:3774–3781
44. Pachl F, Fellenberg K, Wagner C, Kuster B (2012) Ultra-high intra-spectrum mass accuracy enables unambiguous identification of fragment reporter ions in isobaric multiplexed quantitative proteomics. Proteomics 12:1328–1332
45. Ow SY, Salim M, Noirel J, Evans C, Wright PC (2011) Minimising iTRAQ ratio compression through understanding LC-MS elution dependence and high-resolution HILIC fractionation. Proteomics 11:2341–2346
46. Dayon L, Pasquarello C, Hoogland C, Sanchez JC, Scherl A (2010) Combining low- and high-energy tandem mass spectra for optimized peptide quantification with isobaric tags. J Proteomics 73:769–777
47. Kocher T, Pichler P, Schutzbier M, Stingl C, Kaul A, Teucher N, Hasenfuss G, Penninger JM, Mechtler K (2009) High precision quantitative proteomics using iTRAQ on an LTQ Orbitrap: a new mass spectrometric method combining the benefits of all. J Proteome Res 8:4743–4752
48. Michalski A, Damoc E, Lange O, Denisov E, Nolting D, Muller M, Viner R, Schwartz J, Remes P, Belford M, Dunyach JJ, Cox J, Horning S, Mann M, Makarov A (2012) Ultra high resolution linear ion trap orbitrap mass spectrometer (orbitrap elite) facilitates top down LC MS/MS and versatile peptide fragmentation modes. Mol Cell Proteomics 11(O111):013698
49. Ow SY, Cardona T, Taton A, Magnuson A, Lindblad P, Stensjo K, Wright PC (2008) Quantitative shotgun proteomics of enriched heterocysts from Nostoc sp. PCC 7120 using 8-plex isobaric peptide tags. J Proteome Res 7:1615–1628
50. Engmann O, Campbell J, Ward M, Giese KP, Thompson AJ (2010) Comparison of a protein-level and peptide-level labeling strategy for quantitative proteomics of synaptosomes using isobaric tags. J Proteome Res 9:2725–2733
51. Dayon L, Hainard A, Licker V, Turck N, Kuhn K, Hochstrasser DF, Burkhard PR, Sanchez JC (2008) Relative quantification of proteins in human cerebrospinal fluids by MS/MS using

6-plex isobaric tags. Anal Chem 80:2921–2931

52. Dayon L, Turck N, Kienle S, Schulz-Knappe P, Hochstrasser DF, Scherl A, Sanchez JC (2010) Isobaric tagging-based selection and quantitation of cerebrospinal fluid tryptic peptides with reporter calibration curves. Anal Chem 82:848–858
53. Ross PL, Huang YN, Marchese JN, Williamson B, Parker K, Hattan S, Khainovski N, Pillai S, Dey S, Daniels S, Purkayastha S, Juhasz P, Martin S, Bartlet-Jones M, He F, Jacobson A, Pappin DJ (2004) Multiplexed protein quantitation in *Saccharomyces cerevisiae* using amine-reactive isobaric tagging reagents. Mol Cell Proteomics 3:1154–1169
54. Ow SY, Salim M, Noirel J, Evans C, Rehman I, Wright PC (2009) iTRAQ underestimation in simple and complex mixtures: "the good, the bad and the ugly". J Proteome Res 8:5347–5355
55. Addona TA, Abbatiello SE, Schilling B, Skates SJ, Mani DR, Bunk DM, Spiegelman CH, Zimmerman LJ, Ham A-JL, Keshishian H, Hall SC, Allen S, Blackman RK, Borchers CH, Buck C, Cardasis HL, Cusack MP, Dodder NG, Gibson BW, Held JM, Hiltke T, Jackson A, Johansen EB, Kinsinger CR, Li J, Mesri M, Neubert TA, Niles RK, Pulsipher TC, Ransohoff D, Rodriguez H, Rudnick PA, Smith D, Tabb DL, Tegeler TJ, Variyath AM, Vega-Montoto LJ, Wahlander A, Waldemarson S, Wang M, Whiteaker JR, Zhao L, Anderson NL, Fisher SJ, Liebler DC, Paulovich AG, Regnier FE, Tempst P, Carr SA (2009) Multi-site assessment of the precision and reproducibility of multiple reaction monitoring-based measurements of proteins in plasma. Nat Biotechnol 27:633–641
56. Fortin T, Salvador A, Charrier JP, Lenz C, Bettsworth F, Lacoux X, Choquet-Kastylevsky G, Lemoine J (2009) Multiple reaction monitoring cubed for protein quantification at the low nanogram/milliliter level in nondepleted human serum. Anal Chem 81:9343–9352
57. Keshishian H, Addona T, Burgess M, Mani DR, Shi X, Kuhn E, Sabatine MS, Gerszten RE, Carr SA (2009) Quantification of cardiovascular biomarkers in patient plasma by targeted mass spectrometry and stable isotope dilution. Mol Cell Proteomics 8:2339–2349
58. Anderson NL, Anderson NG (2002) The human plasma proteome: history, character, and diagnostic prospects. Mol Cell Proteomics 1:845–867
59. Anderson L, Hunter CL (2006) Quantitative mass spectrometric multiple reaction monitoring assays for major plasma proteins. Mol Cell Proteomics 5:573–588
60. Keshishian H, Addona T, Burgess M, Kuhn E, Carr SA (2007) Quantitative, multiplexed assays for low abundance proteins in plasma by targeted mass spectrometry and stable isotope dilution. Mol Cell Proteomics 6:2212–2229
61. Anderson NL, Anderson NG, Haines LR, Hardie DB, Olafson RW, Pearson TW (2004) Mass spectrometric quantitation of peptides and proteins using stable isotope standards and capture by anti-peptide antibodies (SISCAPA). J Proteome Res 3:235–244
62. Kuhn E, Addona T, Keshishian H, Burgess M, Mani DR, Lee RT, Sabatine MS, Gerszten RE, Carr SA (2009) Developing multiplexed assays for troponin I and interleukin-33 in plasma by peptide immunoaffinity enrichment and targeted mass spectrometry. Clin Chem 55:1108–1117
63. Whiteaker JR, Zhao L, Anderson L, Paulovich AG (2010) An automated and multiplexed method for high throughput peptide immunoaffinity enrichment and multiple reaction monitoring mass spectrometry-based quantification of protein biomarkers. Mol Cell Proteomics 9:184–196
64. Schoenherr RM, Zhao L, Whiteaker JR, Feng LC, Li L, Liu L, Liu X, Paulovich AG (2010) Automated screening of monoclonal antibodies for SISCAPA assays using a magnetic bead processor and liquid chromatography-selected reaction monitoring-mass spectrometry. J Immunol Methods 353:49–61
65. Lange V, Picotti P, Domon B, Aebersold R (2008) Selected reaction monitoring for quantitative proteomics: a tutorial. Mol Syst Biol 4:222
66. Picotti P, Lam H, Campbell D, Deutsch EW, Mirzaei H, Ranish J, Domon B, Aebersold R (2008) A database of mass spectrometric assays for the yeast proteome. Nat Methods 5:913–914
67. Picotti P, Rinner O, Stallmach R, Dautel F, Farrah T, Domon B, Wenschuh H, Aebersold R (2010) High-throughput generation of selected reaction-monitoring assays for proteins and proteomes. Nat Methods 7:43–46
68. Wilm M, Mann M (1996) Analytical properties of the nanoelectrospray ion source. Anal Chem 68:1–8
69. Craig R, Beavis RC (2004) TANDEM: matching proteins with tandem mass spectra. Bioinformatics 20:1466–1467
70. Colinge J, Masselot A, Giron M, Dessingy T, Magnin J (2003) OLAV: towards high-

throughput tandem mass spectrometry data identification. Proteomics 3:1454–1463

71. Boja ES, Phillips D, French SA, Harris RA, Balaban RS (2009) Quantitative mitochondrial phosphoproteomics using iTRAQ on an LTQ-Orbitrap with high energy collision dissociation. J Proteome Res 8:4665–4675
72. Phanstiel D, Unwin R, McAlister GC, Coon JJ (2009) Peptide quantification using 8-plex isobaric tags and electron transfer dissociation tandem mass spectrometry. Anal Chem 81:1693–1698
73. Phanstiel D, Zhang Y, Marto JA, Coon JJ (2008) Peptide and protein quantification using iTRAQ with electron transfer dissociation. J Am Soc Mass Spectrom 19:1255–1262
74. Bantscheff M, Boesche M, Eberhard D, Matthieson T, Sweetman G, Kuster B (2008) Robust and sensitive iTRAQ quantification on an LTQ Orbitrap mass spectrometer. Mol Cell Proteomics 7:1702–1713
75. Savitski MM, Fischer F, Mathieson T, Sweetman G, Lang M, Bantscheff M (2010) Targeted data acquisition for improved reproducibility and robustness of proteomic mass spectrometry assays. J Am Soc Mass Spectrom 21:1668–1679

Chapter 4

Development of MRM-Based Assays for the Absolute Quantitation of Plasma Proteins

Michael A. Kuzyk, Carol E. Parker, Dominik Domanski, and Christoph H. Borchers

Abstract

Multiple reaction monitoring (MRM), sometimes called selected reaction monitoring (SRM), is a directed tandem mass spectrometric technique performed on to triple quadrupole mass spectrometers. MRM assays can be used to sensitively and specifically quantify proteins based on peptides that are *specific* to the target protein. Stable-isotope-labeled standard peptide analogues (SIS peptides) of target peptides are added to enzymatic digests of samples, and quantified along with the native peptides during MRM analysis. Monitoring of the intact peptide and a collision-induced fragment of this peptide (an ion pair) can be used to provide information on the absolute peptide concentration of the peptide in the sample and, by inference, the concentration of the intact protein. This technique provides high specificity by selecting for biophysical parameters that are unique to the target peptides: (1) the molecular weight of the peptide, (2) the generation of a specific fragment from the peptide, and (3) the HPLC retention time during LC/MRM-MS analysis. MRM is a highly sensitive technique that has been shown to be capable of detecting attomole levels of target peptides in complex samples such as tryptic digests of human plasma. This chapter provides a detailed description of how to develop and use an MRM protein assay. It includes sections on the critical "first step" of selecting the target peptides, as well as optimization of MRM acquisition parameters for maximum sensitivity of the ion pairs that will be used in the final method, and characterization of the final MRM assay.

Key words MRM, Multiple reaction monitoring, SRM, Selected reaction monitoring, Mass spectrometry, Quantitation, Internal standards, Stable isotope labeling, Assay development, Plasma, Diagnostics, SIS peptides

1 Introduction

Sensitive and accurate quantitation of proteins by mass spectrometry is conducted by measuring the concentrations of proteolytic and proteotypic peptides, which act as molecular surrogates of the corresponding intact proteins. The use of stable isotopes in quantitative proteomic workflows has had a great impact on improving the quality and reproducibility of MS-based protein quantitation. *Untargeted* MS-based quantitation workflows rely on exhaustive

Helena Bäckvall and Janne Lehtiö (eds.), *The Low Molecular Weight Proteome: Methods and Protocols*, Methods in Molecular Biology, vol. 1023, DOI 10.1007/978-1-4614-7209-4_4, © Springer Science+Business Media New York 2013

sample pre-fractionation methods, such as multidimensional chromatography, which can be performed at both the protein and peptide levels. These techniques have the goal of detecting changes in the expression levels of as many proteins as possible, in an unbiased manner, over a wide dynamic range [1]. These workflows are needed for the "discovery" phase, but are often too expensive in terms of time and reagent costs to be used in the subsequent "verification" or "validation" steps of a biomarker project, or in the ultimate clinical assay, where a large number of samples must be analyzed. Multiple reaction monitoring (MRM) is a tandem MS (MS/MS) technique performed on to triple quadrupole MS instrumentation that is capable of rapid, sensitive, and specific quantitation of *targeted* analytes in highly complex samples [2].

As a targeted method MRM requires knowledge of the molecular weight of an analyte and its fragmentation behavior under collision-induced dissociation (CID) conditions. By combining carefully chosen of peptides and selection of MRM precursor and fragment ion pairs with the use of stable isotope-labeled standard (SIS) peptides, MRM can be used to determine, highly specifically and reproducibly, the absolute concentrations of peptides in the digest. The concentrations of the peptides can be used to infer the concentrations of the parent proteins. There are several critical steps involved in the development of an MRM assay and they all rely on a common set of protocols. This chapter has been divided into two sections: the first section covers the steps required to develop an MRM protein assay. These include the selection of the proteotypic peptides that will "represent" the protein in the final assay, the synthesis of isotopically labeled analogues of these peptides to be used as internal standards, and the optimization of MRM acquisition parameters so that sensitive and accurate quantitation can be performed without interference from other components in the sample. Finally, MRM assays must be characterized to determine the technical reproducibility and concentration range of the linear response over which quantitation can be performed. We have included in this section the methods for synthesis of the SIS peptides, in case the required SIS peptides are not commercially available.

The second section describes all of the methods required for preparation and mass spectrometric analysis of plasma samples by MRM. Plasma is a readily available biological fluid, and is easily collected, which makes it attractive for biomarker studies. However, plasma is an extremely complex (10^{10}-fold protein concentration range) and highly proteinaceous sample matrix (60–80 mg/mL) which is frequently used for both discovery proteomics and for quantitative proteomics of disease biomarkers [3]. By employing a directed quantitation method like MRM, plasma proteins can be specifically targeted and quantified even though ~90 % of the total protein by weight in plasma can be attributed to ten high-abundance proteins [4]. When used with isotopically labeled synthetic peptide standards and either depletion of high-abundance proteins or affinity enrichment of target proteins, MRM analysis is capable of sensitive (attomole level) and absolute determination of peptide concentrations across a concentration range of 10^3–10^4 [5–10]. We used these methods for our paper on the "top 45" proteins in plasma [11], and have since been using these protocols to develop MRM methods for the absolute quantitation of more than 90 proteins in plasma. The protocols given here do not include depletion or enrichment steps, which could be added to the sample preparation step if needed. By changing the sample preparation steps, these methods could also be used for developing MRM-based methods for other types of samples, such as cells or tissues.

2 Materials

Please note that other reagents and suppliers may be substituted and may give equally good results, but we have had no experience with their use. In each case where a particular manufacturer is listed, please recognize that the phrase "or the equivalent" is implied.

2.1 Confirmation of Proteotypic Peptide Detection

1. Ammonium bicarbonate (ABC).
2. Dithiothreitol (DTT).
3. 37 °C oven.
4. Iodoacetamide.
5. Trypsin.
6. C_{18} ZipTip (Millipore, or the equivalent).

2.2 Synthesis of Stable Isotope-Labeled Standard (SIS) Peptides

1. Water, LC/MS grade.
2. Trifluoroacetic acid (TFA).
3. L-Lysine-α-*N*-Fmoc, ε-*N*-t-Boc [$^{13}C_6$] (98 % isotopic enrichment) (Cambridge Isotope Laboratories, Andover, MA), or the equivalent.
4. L-Arginine-*N*-Fmoc, PMC [$^{13}C_6$][$^{15}N_4$] (98 % isotopic enrichment) from Cambridge Isotope Laboratories (Andover, MA), or the equivalent.
5. TentaGel R resin from Rapp Polymere (Tubingen, Germany), or the equivalent.
6. Piperidine.
7. ^{1}H-Benzotriazolium 1-[bis(dimethylamino)methylene]-5-chloro-hexafluorophosphate (1),3-oxide (HCTU) (Protein Technologies, Tucson, AZ), or the equivalent.
8. Triisopropylsilane.

2.3 HPLC Purification of Isotopically Labeled SIS Peptides

1. Diethyl ether.
2. Trifluoroacetic acid (TFA).
3. Water, LC/MS grade.
4. Acetonitrile, LC/MS grade.
5. Onyx Monolithic C18 column, 10 mm × 100 mm, 13 nm mesopores, 2 μm macropores (Phenomenex, Torrance, CA), or the equivalent.
6. Harvard PicoPlus 11 Syringe Pump, or the equivalent.
7. Gas-tight syringe (100 μL).

2.4 Empirical Optimization of MRM Acquisition Parameters

1. Acetonitrile.
2. Formic acid.
3. 100 μL gas-tight syringe.
4. AB/MDS Sciex 4000 QTRAP (or equivalent triple quadrupole).

2.5 Interference Detection and Verification of Co-elution

1. SIS peptides.
2. Formic acid.
3. Polypropylene autosampler vials.

2.6 Optimizing the SIS Peptide Spiking Concentration

1. SIS peptides.
2. Formic acid.
3. Solid-phase extraction (SPE) cartridges.
4. Water, LC/MS grade.

2.7 Sample Preparation for MRM Analysis

2.7.1 Collection of EDTA-Plasma

1. BD P100 Blood Collection Kit containing K_2·EDTA and proprietary protein stabilizers (Becton Dickinson, Franklin Lakes, NJ).
2. Screwtop microtubes, 2 mL.

2.7.2 Tryptic Digestion of Plasma

1. Sodium deoxycholate.
2. Ammonium bicarbonate.
3. Bond-Breaker tris(2-carboxyethyl)phosphine (TCEP) Neutral pH (Thermo Scientific, Rockford, IL).
4. Iodoacetamide (Fluka Sigma-Aldrich, Steinheim, Germany).
5. Trypsin, modified porcine sequencing-grade (Promega, Madison, WI).
6. Formic acid (0.1 % v/v) (Fluka Sigma-Aldrich, Steinheim, Germany).
7. Screwtop microtubes, 2 mL.
8. Pipettor, 1,000 μL capacity accurate to 1 μL.

2.8 Solid-Phase Extraction (SPE)

1. Methanol, LC/MS grade.
2. Oasis reversed-phase 1 cc, 10 mg HLB cartridges (Waters, Milford, MA).

2.9 LC/MRM-MS Analysis of Tryptic Plasma Digests

1. AB Sciex QTrap 4000, or other triple quadrupole mass spectrometer.
2. IntegraFrit fused silica column tubing, 75 μm ID × 360 μm OD, 15 cm (New Objective, Woburn, MA).
3. Reversed-phase packing material, Magic C18AQ, 5 μm ID particles, 100 Å pore size (Michrom, Auburn, CA).
4. All HPLC capillaries used are fused silica tubing (20 μm ID × 360 μm OD) (Polymicro Technologies, Phoenix, AZ).
5. Fused silica, uncoated emitter tips, 20 μm inner diameter, 10 μm tip (New Objectives, Woburn, MA).
6. Water, LC/MS grade.
7. HPLC Solvent A: 2 % v/v acetonitrile, LC/MS grade, 0.1 % v/v formic acid (Fluka Sigma-Aldrich, Steinheim, Germany).
8. HPLC Solvent B: 98 % v/v acetonitrile, LC/MS grade, 0.1 % v/v formic acid.

3 Methods

There are several sections to these protocols. If you are following an already developed method, skip to Subheading 3.3 for tuning and optimization of the method on your particular mass spectrometer. "Absolute" quantitation, as used here, still requires comparison of the signal from a component in the sample with that of a standard of known concentration. Isotopically labeled standard peptides are used in MRM analysis because they can be distinguished in mass from the analogous endogenous peptide. If the SIS peptides are already purified and commercially available, skip to the appropriate part of Subheading 3.4. If the SIS peptides are commercially available, purified, and the accurate concentrations are known, skip to Subheading 3.7 for sample preparation and analysis.

3.1 MRM Assay Development

Development of an MRM assay is often an iterative process that involves the selection, synthesis, and characterization of multiple peptides that will act as surrogates for the protein. An overview of the workflow used to develop an MRM method is presented in Fig. 1 with each step in the process described below.

3.1.1 Selection of Target Peptides

Since targeted-proteomics approaches such as MRM require knowledge of a protein's sequence, peptide selection is initially a

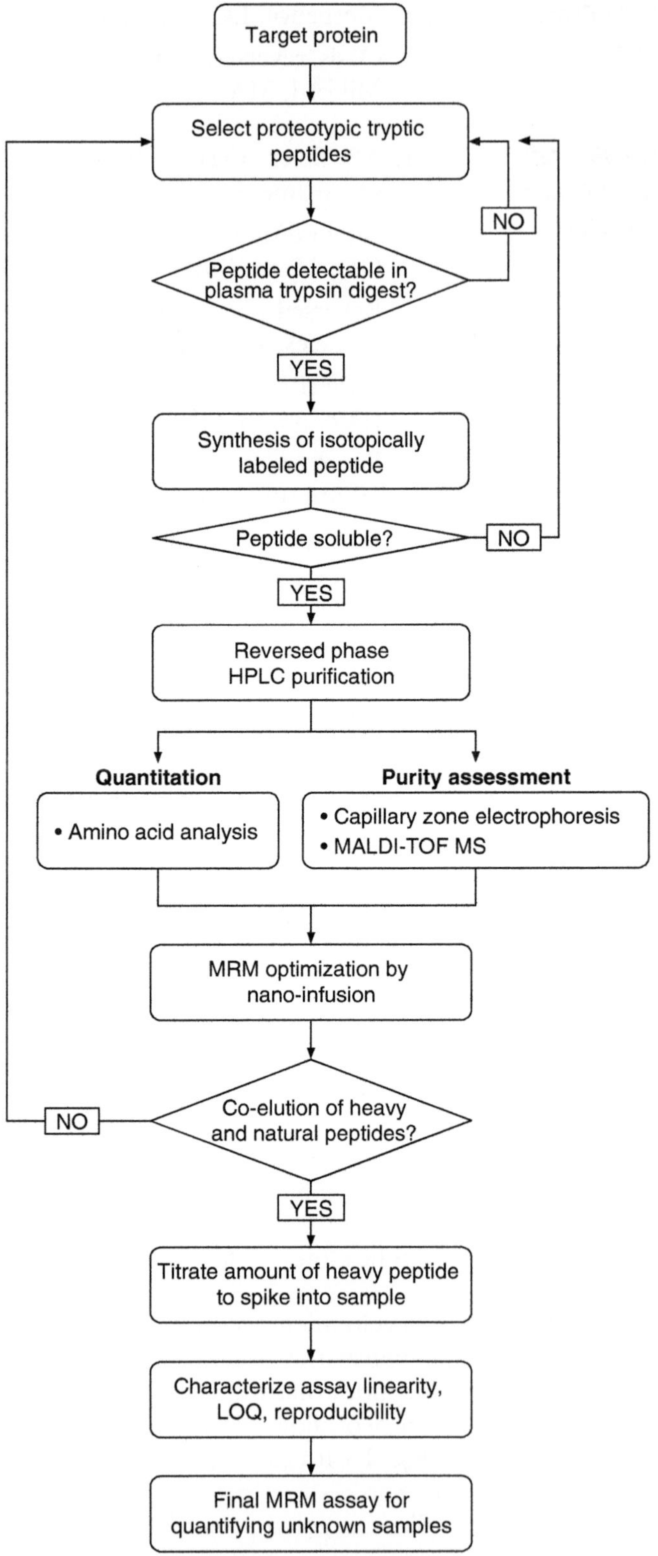

Fig. 1 Strategy for generation of highly sensitive and specific MRM protein assays. Adapted from ref. 11, with permission

bioinformatics exercise. Depending on the biological system—for this chapter we are using plasma as the example—mouse, rat, or human, there are several useful databases of MS/MS data that are publically accessible on the internet. These resources can be used to identify proteotypic peptides, and peptides that are frequently observed in tryptic digests of a target protein: Peptide Atlas (http://www.peptideatlas.org/) is a well-annotated database hosted by the Institute for Systems Biology with specific builds for mouse and human plasma, and the Global Proteome Machine Database [12] (http://www.gpmdb.org/) is an automatically generated repository of MS/MS spectra anonymously deposited by researchers using the X!Tandem MS/MS spectrum modeler for protein identification.

There is currently a great deal of interest in the bioinformatics community in developing software to assist in MRM method development. Some of these software packages help in determining proteotypic peptides [13], while others help predict MRM ion pairs. These include Skyline SRM/MRM builder [14–16], AB's MRMpilot [17], MaRiMba[18], MRMaid [19], and MRMer [20]. There are also several libraries of MRM ion pairs currently being developed from literature data. These include new libraries of MRM pairs for target proteins such as cancer biomarker peptide libraries [21], yeast [22], as well as other libraries generated from data in the literature or the Global Proteome Machine Database [12].

Peptide selection rules for MRM assay development are as follows:

1. The amino acid sequence of a selected peptide must be *completely unique* to the target protein within the target biological system. PeptideAtlas actually indicates multiple genome locations at the tryptic peptide sequence level. A BLASTp analysis, limited to the taxonomy of the biological system, is the ultimate confirmation of sequence uniqueness. Attention must be paid to Leu/Ile which are isobaric by most mass spectrometric techniques, but are not homologous by BLASTp analysis. Additionally, the preceding amino acid required for the cleavage site should be used to extend the sequence when conducting homology searches. For example, the tryptic peptide THGFLR may be identical to another 6-mer, but when we consider that the sequence must be [R]THGFLR or [K]THGFLR in order to generate a tryptic peptide, we should "virtually" extend the length of this peptide when considering sequence homology to other proteins.
2. Peptides must not contain missed tryptic cleavage sites (i.e., internal K or R residues), since variations in digestion efficiency might produce different forms of the same peptide.
3. Peptide length should be >5 amino acids and <25 amino acids, for specificity and synthesis efficiency, respectively. Also, peptides

that are too short may have protonated molecular ions that are within the region that is masked by the solvent in LC/MRM-MS analyses. Peptides that are too long may ionize with poor efficiency and fragment poorly in MS/MS.

4. Peptide sequences should not contain the following (*see* **Note 1**):
   (a) Asp-Pro (D-P) in the sequnce. Asp is particularly susceptible to dehydration to form a cyclic imide intermediate in the presence of Pro, which can lead to cleavage of the peptide chain in acidic conditions.
   (b) Cysteine (C), which causes peptides to form insoluble aggregates through the formation of disulfide bonds. Reduction/alkylation, if incomplete, may lead to multiple forms, thus dividing the signal between several different ions.
   (c) Methionine (M), which is highly susceptible to oxidation which leads to the formation of sulfoxide (+16 Da mass shift) and sulfone derivatives (+32 Da mass shift). This "splits" the peptide signal over several different molecular weights, thus reducing the sensitivity of each signal, which could also ultimately reduce the accuracy of the analysis.
   (d) N-terminal glutamine (Q), which cyclizes to form pyroglutamate.
   (e) Tryptophan (W), which is also susceptible to mono- and di-oxidation (+16 and/or +32 Da mass shifts) (*see* **Note 2**).
   (f) When data is available (and assuming that determination of posttranslational modification or peptide modification is not the goal of the assay), peptides should not contain residues known to be post-translationally modified. As with oxidation, this would mean that several different forms of the same peptide would have to be monitored.
   (g) Peptides should also not contain known single nucleotide polymorphisms (SNPs) that encode amino acid substitutions. Again, if SNPs exist, this would mean that several different forms of the same peptide would have to be monitored in order to accurately determine the concentration of all forms of the target protein.

3.1.2 Confirmation of Proteotypic Peptide Detection

If a recombinant source of the target protein is available, it is useful to confirm observation of the selected target peptides in an actual tryptic digest of the target protein prior to investing in the synthesis of isotopically labeled peptides. Obviously, the sensitivity of the target peptide is going to be an important factor in the overall sensitivity of the assay. Because peptide sensitivities are dependent on ionization techniques, if ESI is going to be used for the LC/MRM-MS analysis, then the sensitivity of the peptide should be confirmed by ESI.

1. Add 25 mM ammonium bicarbonate (ABC) to ~10 μg of protein to a final volume of 100 μL.
2. Reduce the sample by adding 5 μL of DTT (200 mM DTT, 25 mM ABC).
3. Incubate at 37 °C for 30 min.
4. Alkylate free sulfhydryl groups with 20 μL of iodoacetamide (200 mM iodoacetamide, 25 mM ABC).
5. Incubate at 37 °C for 30 min in darkness.
6. Add 10 μL of trypsin (0.1 μg/μL, 25 mM ABC) and incubate at 37 °C for 16 h.
7. Desalt and concentrate the sample using a C_{18} ZipTip and analyze ~1 pmol of digested protein by data-dependent LC/MRM-MS analysis to confirm that proteotypic peptides selected by bioinformatics analysis are observed in an actual tryptic digest of the protein.

3.2 Stable-Isotope-Labeled Peptides

Once the sensitivities and specificities of the peptides have been determined, their stable-isotope-labeled analogues can be synthesized or purchased. It is important to remember that the accuracy of your final assay will be based on the signals from these peptides. For this reason, it is critical that these standards are purified so that their concentrations can be accurately determined.

3.2.1 Synthesis of Stable-Isotope-Labeled Peptides

If stable-isotope-labeled peptides are commercially available for this assay, skip to the appropriate section below (Subheading 3.3.1 or 3.3.2).

The FMOC procedure used for synthesis of the stable-isotope-labeled standard (SIS) peptides is outlined in detail in Kuzyk et al. [11]. Although peptide synthesis is frequently performed as a service, we find it worthwhile to synthesize peptides in house to retain greater control over the purity of the final product. Here we only provide the salient details involved in the synthesis of stable isotope-labeled standard (SIS) peptides since we recognize that peptide synthesis equipment is not commonplace in many laboratories. We exclusively use [^{13}C][^{15}N]-labeled peptides, and not deuterated analogues, because in reversed-phase HPLC, deuterated and non-deuterated peptides do not exactly co-chromatograph and it is easier to align isotopically labeled and native peptides which exactly co-elute and where this is not an issue.

1. A peptide synthesizer can be used to synthesize up to 96 peptides in parallel using FMOC chemistry at a 5 μmol synthesis scale.
2. A resin is used for peptide synthesis and we usually use a resin conjugated to isotopically labeled Arg or Lys (*see* **Note 3**).

3. [$^{13}C_6$]Lys (98 % isotopic enrichment) and [$^{13}C_6$][$^{15}N_4$]Arg (98 % isotopic enrichment) can be purchased and sent out for cross-linking to the resin (*see* **Note 4**).
4. Double-couple subsequent (*see* **Note 5**) amino acid residues (100 mM) and use 20 % piperidine as the deprotector and ^{1}H-Benzotriazolium 1-[bis(dimethylamino)methylene]-5-chloro, hexafluorophosphate (1),3-oxide (HCTU) as the activator.
5. Cleave completed peptide off the resin with 95:2.5:2.5 v/v/v TFA:water:triisopropylsilane.

3.3 Purification and Determination of Stable-Isotope-Labeled Peptides

3.3.1 Purification

1. Remove the peptides from the synthesizer and evaporate the TFA in a fume hood using a stream of nitrogen to reduce volume ~fourfold.
2. Add 30 mL of cold ether to precipitate the peptides.
3. Centrifuge at 3,000 × *g* for 15 min and decant supernatant.
4. Resolubilize the peptides in 0.1 % TFA. Following cleavage off of the resin and precipitation, some peptides are unable to be resolubilized. When this occurs, an alternate proteotypic peptide must be selected for synthesis (i.e., go back to Subheading 3.2.1) (*see* **Note 6**).
5. Purify the peptides by reversed-phase HPLC, using a C_{18} column with a linear gradient of 0.1 % TFA in water (v/v) and 0.85 % TFA in acetonitrile (v/v) at a flow rate of 5 mL/min over 25 min.
6. Injection volume depends on the volume used during resolubilization. However, these should typically not exceed 2 mL.
7. Collect fractions by time, every 10 s.
8. Based on the UV chromatogram, spot fractions of interest onto MALDI plates for analysis (*see* **Note** 7).
9. Pool fractions that predominantly contain the target peptide as determined by MALDI-TOF MS analysis.
10. Lyophilize to dryness.

3.3.2 Concentration and Purity Determination of SIS Peptides

To enable concentration determination by MRM analysis, the concentration of a SIS peptide needs to be accurately determined. If a SIS peptide is purchased from a commercial supplier, you should receive a certificate of analysis that states the concentration and purity of the synthetic peptide and the methods used to determine these values. Acid hydrolysis of peptides followed by amino acid analysis (AAA) is commonly used to determine the peptide concentration; however this concentration can be affected by the presence of incomplete or partial synthesis products that co-purified with the target peptide during HPLC.

To correct the peptide concentration obtained by AAA so that it accurately reflects the concentration of the intact target SIS peptide, the percent purity of a peptide must be determined. To determine percent purity, either RP-HPLC analysis or capillary zone electrophoresis with ultraviolet detection (190–210 nm) can be used. Percent purity is calculated by integrating the total area of all peaks observed during the purity analysis, and determining what percentage of the total peak area is comprised by the most abundant peak (i.e., the target peptide). The AAA-derived concentration is then adjusted by multiplying it by the percent purity value.

3.4 Optimization of the MRM Method

3.4.1 Empirical Selection of Optimum Charge State

Following the synthesis of the SIS peptides, there are several parameters that need to be optimized to maximize the sensitivity of detection of the mass spectrometric portion of the assay. Since an isotopically labeled peptide is chemically identical to its endogenous form, the synthetic SIS peptides can be used to tune MRM acquisition parameters.

Empirical optimization of all MRM acquisition parameters is conducted by infusing a solution of pure synthetic SIS peptide at nL/min flow rates using a syringe pump. The first step focuses on determining the dominant precursor ion charge state of a peptide, by ramping the declustering potential (DP) voltage to maximize the transfer efficiency of the precursor ion into the mass spectrometer. Each charge state of a peptide will reach a maximum signal intensity at a specific DP voltage (Fig. 2).

1. Dilute SIS peptides to 1 pmol/μL (1 mM) in 30 % v/v acetonitrile, 0.1 % v/v formic acid.
2. Infuse the peptide solution using a 100 μL gas-tight syringe at a flow rate of 300 nL/min using a syringe pump.
3. Using an AB/MDS Sciex 4000 QTRAP (or equivalent triple quadrupole), set up a quadrupole 1 (Q1) scan with 5-Da-wide mass ranges centered on the predicted *m/z* values of the double and triple charge states of the target peptide.
4. Acquire data while ramping declustering potential (DP) from 0 to 120 V in 2 V increments.
5. Note which charge state reaches the highest signal intensity and the DP voltage required to obtain that signal as these parameters will be used to create MRM ion pairs in the next section.

3.4.2 Optimization of Collision Energy

In an MRM assay, the sensitivity of an MRM ion pair is directly related to the signal intensity of the fragment ion that is transmitted through quadrupole 3 (Q3) and allowed to strike the detector. Although low-energy collision-induced dissociation (CID) preferentially fragments peptides at the peptide bond, it is not possible to predict which fragment ions will generate the highest signals. Additionally, the collision energy voltage in the mass spectrometer can be varied, which dramatically affects the fragmentation pattern of a peptide and the relative intensities of the fragment ions.

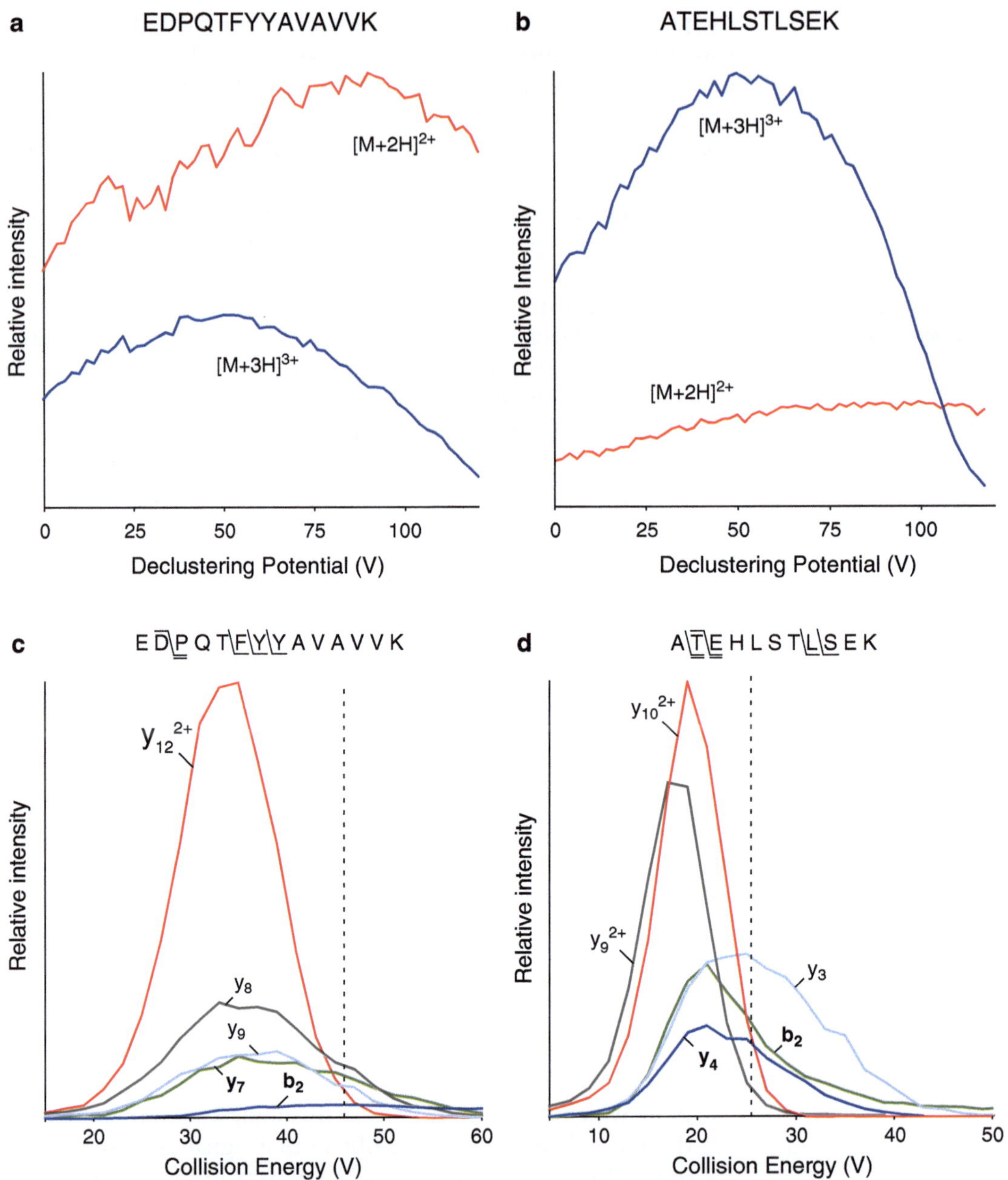

Fig. 2 (**a, b**) Optimization of declustering potential and (**c, d**) collision energy for two peptides from transferrin (*left*) and apolipoprotein A-1 (*right*). Reprinted from ref. 11, with permission. The labeling convention for peptide fragment ions is the Roepstorff nomenclature [29]. In this system, b-ions include the N terminus while y-ions include the C terminus of the peptide

Publically available MS/MS databases such as PeptideAtlas can be used to select fragment ions. However, the MS/MS spectra in these databases are acquired with the goal of peptide *identification* which requires CID conditions that generate rich fragmentation spectra representing the entire amino acid sequence of a peptide.

When optimizing MRM parameters for a peptide the intention is to fragment a peptide into only a *few* high-intensity fragment ions. Although there are software packages, for example MIDAS™ [23], which can be used to predict MRM Q1/Q3 ion pairs in silico, we have previously reported that empirically optimized MRM parameters increased the MRM signal intensity of 45 peptides by an average of 11.4-fold compared to those predicted by MIDAS [11].

1. Switch the MS to an MRM scanning mode and create MRM ion pairs using the dominant precursor ion charge state as the mass value for Q1 and the corresponding optimized DP voltage.
2. For the Q3 portion of the MRM ion pairs use the predicted m/z values of each predicted b-, a-, y-, and y^{2+}-ion series (*see* Fig. 2c. for fragment ion nomenclature). These can be calculated using publically available free software tools (*see* **Notes 8** and **9**).
3. For each MRM ion pair use a dwell time of 100 ms. Since this optimization step is conducted with the target peptide being continually syringe infused into the mass spectrometer, there is no need to limit the number of fragment ions that are monitored.
4. Collect MRM data while ramping the collision energy (CE) voltage (V) from 5 to 120 V in 2 V increments.
5. Generate extracted ion chromatograms for each MRM ion pair and identify the five highest intensity MRM ion pairs and their corresponding CE V to use to create an optimized MRM acquisition method (Fig. 2d).

3.5 Interference Detection and Verification of Co-elution

Although MRM ion pairs are highly specific to the target peptide with two stages of mass discrimination, and a third retention time-specific criteria, it is possible for a co-eluting ion species to interfere with the MRM measurement in a sample as complex as a plasma digest. Following optimization of the most sensitive MRM ion pairs for a peptide, it is necessary to determine which MRM ion pairs are free of interferences and can be used for reliable quantitation in the final MRM assay. This analysis can also be combined with confirming the co-elution of SIS and endogenous peptides during the LC/MRM-MS analysis of a real sample spiked with the SIS peptide. This is a necessary step to verify the identity of the observed endogenous (or "light") MRM signal during an LC/MRM-MS analysis. We can achieve both goals by monitoring multiple MRM ion pairs for both the endogenous and SIS peptides, but the pure SIS peptide must first be analyzed alone by LC/MRM-MS, *followed by* the analysis of the SIS and native peptides in the context of the target sample. The relative signal intensities of the MRM ion pairs for both the heavy and endogenous forms of a peptide must remain the same when analyzed in solvent and in a plasma digest to be qualified as free of interferences.

1. Dilute the SIS peptide to a concentration of 100 fmol/μL with 0.1 % v/v formic acid. Transfer 20 μL to a polypropylene autosampler vial for LC/MRM-MS analysis.
2. Create an MRM acquisition method using the five highest intensity MRM Q1/Q3 ion pairs with their optimized DP and CE voltages for the SIS peptide.
3. Since these Q1 and Q3 masses of these ion pairs are specific to the SIS peptide, the masses need to be adjusted to create MRM ion pairs that correspond to the endogenous form of the peptide. However, identical DP and CE voltages should be used for both the endogenous and SIS peptide MRM ion pairs.
4. Inject 1 μL of the SIS peptide (100 fmol) and analyze by LC/MRM-MS using an acquisition method containing MRM ion pairs for both the SIS and the endogenous forms of the peptide. Any signal observed in the XIC trace for the endogenous form of the MRM ion pair can be attributed to isotopic impurities in the SIS peptide (*see* **Note 10**).
5. In total, perform five replicate LC/MRM-MS analyses of pure SIS peptide in solvent (0.1 % formic acid) to determine the relative intensities of the five MRM ion pairs in the absence of a sample matrix.
6. Next, perform five replicate LC/MRM-MS analyses of a sample spiked with SIS peptides that is representative of the type of sample that will be analyzed with the final MRM assay. This sample needs to contain either a detectable amount of the endogenous protein or a detectable amount of the recombinant protein must be spiked in prior to trypsin digestion (*see* **Note 11**).
7. Integrate the peak areas of all MRM ion pairs using an analysis software package such as MultiQuant. Confirm that MRM signals from the SIS and endogenous forms of the peptide co-elute; if they do not co-elute, a new proteotypic peptide must be selected for assay development.
8. At this stage the relative ratio of the MRM ion pair areas to one another (heavy and light) needs to be calculated for each peptide (Table 1).
9. Identify and exclude any MRM ion pairs that generate relative intensities that disagree with the values calculated from analysis of the peptide in solvent alone.

Interference detection is a crucial step used to determine which MRM ion pairs are free from interferences from other components which might be present in the sample. At least three MRM ion pairs free of interferences are needed for quantitative analysis of unknown samples. The highest intensity MRM ion pair, containing a y fragment ion that is free of interferences, should be used for quantitation (*see* **Note 12**). Two additional MRM ion pairs free of

Table 1
Example output from an interference detection experiment

	Average peak area (count per second)			Relative intensity			
Fragment ion	SIS in solvent	SIS in plasma	Natural in plasma	SIS in solvent	SIS in plasma	Natural in plasma	Average peak area ratio
y_{10}^{2+}	5.98 E+06	1.72 E+06	1.65 E+06	1.00	1.00	1.00	0.96
y_9^{2+}	5.12 E+06	1.46 E+06	1.40 E+06	0.86	0.85	0.85	0.96
a_2	3.05 E+06	8.33 E+05	7.97 E+05	0.51	0.48	0.48	0.96
y_3	2.90 E+06	7.90 E+05	7.26 E+05	0.48	0.46	0.44	0.92
b_2	2.62 E+06	7.06 E+05	6.70 E+05	0.44	0.41	0.41	0.95

In this example using the ATEHLSTLSEK peptide of human apolipoprotein A-I, the y_{10}^{2+} MRM ion pair would be selected as the quantifier ion pair, since it is the highest intensity y-ion-containing ion pair that is free of interference. The y_9^{2+}- and a_2- containing ion pairs would be selected as qualifier peaks to include in the analysis of unknown samples to allow the detection of interferences in real biological samples. All three ion pairs generate the same average peak area ratio (endogenous:SIS)

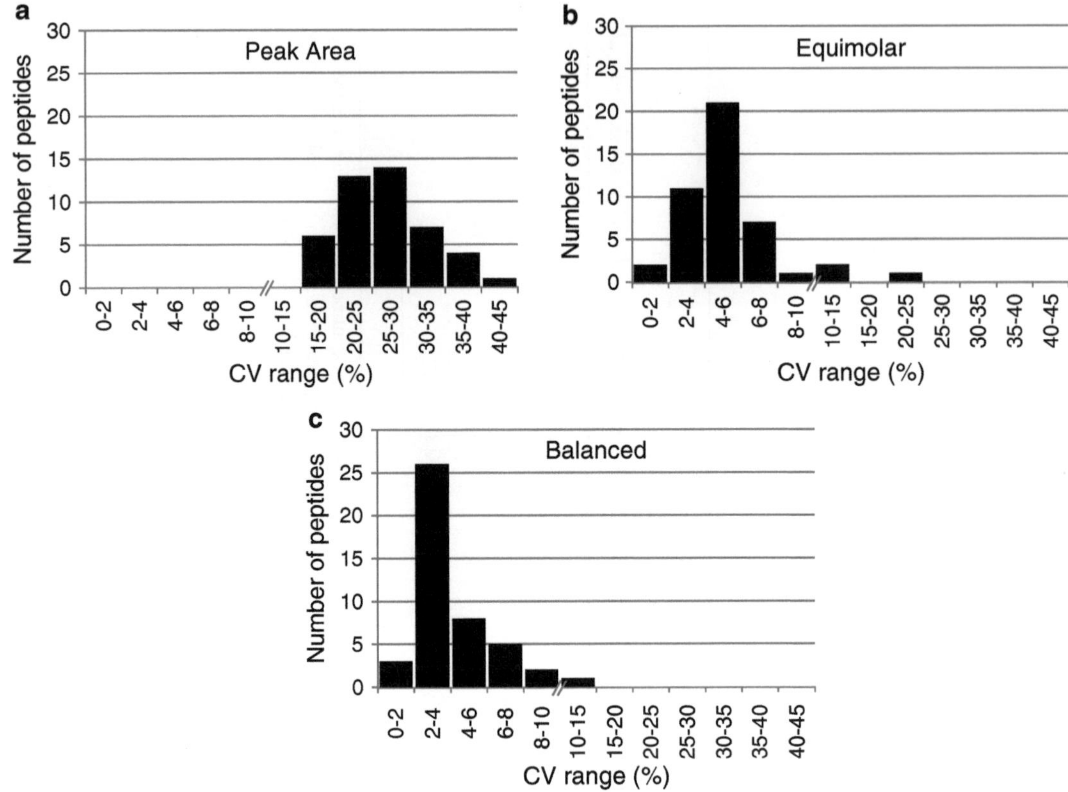

Fig. 3 Analytical reproducibility of MRM-based quantitation. CV frequencies of 45 peptide assays using (**a**) raw peak areas, (**b**) peak area ratios normalized to an equimolar SIS peptide mixture, and (**c**) peak area ratios normalized to a concentration-balanced SIS peptide mixture. Reprinted from ref. 11, with permission

interferences should be selected to be used as qualifier ion pairs. In the analysis of unknown biological samples, "qualifier" ion pairs can be used to compare the relative intensities of endogenous and SIS peptides to those obtained with the standard sample. If an unknown sample contains a co-eluting interference, it can be identified using the relative intensities calculated with the qualifier ion pairs.

These three, interference-free, MRM ion pairs constitute the final MRM assay for the proteotypic peptide.

3.6 Optimizing the SIS Peptide Spiking Concentration

Determining the optimal concentration of a SIS peptide to add to plasma maximizes the linear range of the MRM assay and ensures that the SIS peptide generates a high-quality signal upon analysis, thereby reducing the analytical variation of the assay (Fig. 3).

1. Create a dilution series of your SIS peptide in 0.1 % v/v formic acid that ranges in concentration from 0.7, 3.5, 7, 17.5, 35, and 70 pmol/μL which equates to 10, 50, 100, 250, 500, and 1,000 fmol of SIS peptide on column per analysis (*see* **Note 13**).

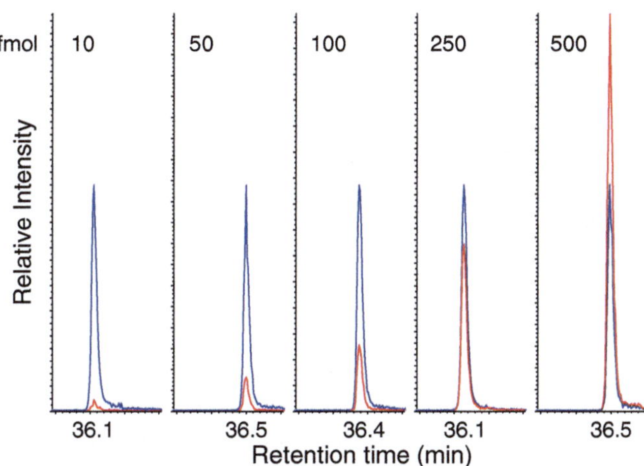

Fig. 4 Titration of the optimal spiking concentration of a SIS peptide for haptoglobin. Extracted ion chromatograms (XICs) are presented for both the endogenous, light peptide (*gray line*) and its co-eluting, heavy SIS peptide (*black line*). In this case, the 35 pmol/μL plasma spiking concentration (500 fmol on column, per analysis) would be chosen for the haptoglobin MRM assay

2. Spike each SIS peptide concentration into a tryptic digest of a plasma sample that contains the target protein at concentrations that will be encountered during quantitation with this assay.
3. Desalt and concentrate the spiked digests by SPE.
4. Rehydrate the six samples and analyze them by LC/MRM-MS.
5. The SIS peptide concentration that generates a peak area ratio (light versus heavy) within tenfold of the endogenous peptide signal (*preferably higher concentration than the analyte*) should be used for all subsequent analyses (Fig. 4).

3.6.1 Calibration Curves and Linear Range of Quantitation

Calibration curves are used in MRM quantitation to determine how the measured ratio of endogenous analyte versus internal standard responds to changes in the actual concentration of the endogenous analyte. This response can be affected by variations in trypsin digestion and performance of the mass spectrometer. Therefore, it is necessary to prepare and analyze a calibration curve to quantify unknown samples in each analytical batch.

The standard method for creating a calibration curve is to add varying amounts of a standard protein to constant amounts of blank sample matrix and internal standards. However, when assaying a protein that is naturally present in all samples, preparation of a blank sample matrix devoid of the target endogenous protein is not feasible. We therefore employ an alternative method, similar to that of Janecki et al. [24], in which a tryptic digest of a standard plasma sample is diluted to create a series of concentrations and mixed with a constant amount of SIS peptide [11, 24].

Table 2
Dilution series used to characterize linear quantitation range of an MRM assay

Sample number	1	2	3	4	5	6	7
Effective plasma volume (nL)	536	107	214	42.9	8.6	1.7	0.34
Amount of SIS peptide (pmol)	0.75	0.75	7.5	7.5	7.5	7.5	7.5
Plasma conc. of SIS peptide (pmol/μL)	1.4	7.0	35	175	875	4,375	21,875

Sample number 3 (*gray shading*) represents the standard ratio of SIS peptide spiked to plasma digest that would be used to obtain 500 fmol on column per analysis

1. Perform a fivefold dilution series of a standard plasma sample tryptic digest to generate a range of endogenous analyte concentrations that span a 15,000-fold concentration range.
2. Spike each sample concentration with a constant amount of SIS peptide. Using the example from Fig. 4, we have determined that the SIS peptide should be spiked into plasma at a concentration of 35 pmol/μL (500 fmol on column). The dilution series of sample versus SIS peptide should be created as outlined in Table 2 (*see* **Note 14**).
3. Prepare enough of each sample concentration for 15 LC/MRM-MS analyses to permit reanalysis if necessary (Table 2).
4. Desalt and concentrate the spiked digests by SPE.
5. Rehydrate the seven samples and conduct five replicate LC/MRM-MS of each sample concentration, ensuring that a constant amount of SIS peptide is injected per analysis. *We recommend analyzing samples in the order of increasing concentration, with blank injections acquired between differing sample concentrations.*
6. Integrate the MRM data using MultiQuant or similar analysis software.

Linear regression analysis of the observed MRM peak area ratios (endogenous:heavy) plotted versus concentration ratios is used to generate calibration curves (Fig. 5a). To permit calculation of concentration ratios, we first need to estimate the standard sample's endogenous concentration of the target analyte if the actual concentration is unknown. To accomplish this, we use the five replicate analyses of sample 3 (Table 2) to calculate the concentration of the endogenous peptide *relative to the known concentration of its SIS peptide* (*see* **Note 15**).

To avoid over-representing the degree of linearity in a calibration curve when defining the linear concentration range of an MRM assay, it is important to combine linear regression analysis with response factor plots to identify analyte concentrations that are not responding in a linear manner [25]. Response factor plots should have a near-zero slope.

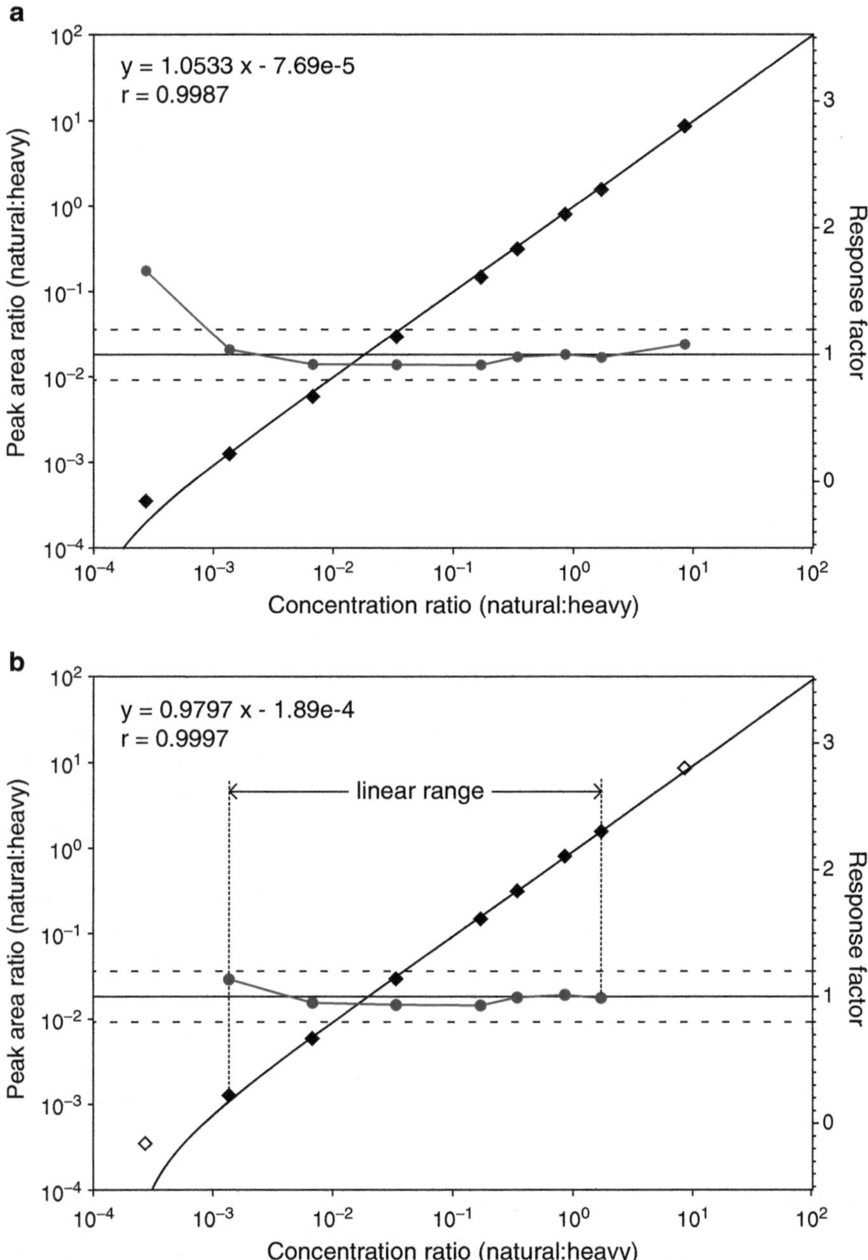

Fig. 5 Calibration curves and response factor plot for apolipoprotein A-I. (**a**) Linear regression analysis (1/*x* weighted) and response factor plot (*horizontal line*) for ApoA-I using all analyte concentrations. (**b**) Final linear regression and response factor plot for ApoA-I illustrating the linear dynamic range of the assay when analyte concentrations that respond non-linearly are excluded (*open diamond*). Reprinted from ref. 11, with permission

7. To create a response factor plot, the known concentration ratio and measured peak area ratio for each sample concentration are used as the x and y values in the linear regression formula ($y = m(x) + b$) to solve for m (slope).

8. If the response factor for a sample concentration differs from the actual slope of the linear regression by greater than 20 %, it is excluded from the next round of linear regression analysis (Fig. 5a).
9. Repeat the linear regression analysis without the excluded values and generate a new response factor plot using the new linear regression formula.
10. Repeat **steps 1–3** until all sample concentrations used in the linear regression have response factors that fall within 80–120 % of the average analyte response (slope) (Fig. 5b) [25]. This ensures that the best linear fit is achieved, thereby improving quantitation accuracy, and accurately representing the linear concentration range of each peptide assay.

3.6.2 Determination of the MRM Assay Lower Limit of Quantitation

The limit of quantitation (LOQ) for an assay is an important measure of performance as it defines the lowest analyte quantity that can be accurately measured, and it is the true measure of the sensitivity of an assay. A signal-to-noise ratio of >10 is frequently used to define LOQ [26, 27]. However, since MRM is an MS/MS scan mode, background noise is extremely low and there is no practical way to accurately measure co-eluting noise in an MRM Q1/Q3 ion pair channel of interest. Therefore, it is advantageous to empirically determine an assay's LOQ, defined as the lowest analyte concentration which can be measured with <20 % coefficient of variance (CV) [25]. The five replicate LC/MRM-MS analyses of each sample concentration used to define assay linear range can be used to estimate the LOQs for an MRM assay. However, to robustly define an LOQ for an MRM assay, it is necessary to characterize the technical reproducibility of quantitation at this concentration. The definition of an MRM assay LOQ is laboratory specific; however, many labs will define an LOQ concentration based on the reproducibility of an assay performed on a standard sample in replicate on different days (six technical sample replicates with a calibration curve, generated on 3 separate days).

3.6.3 Analyzing Unknown Samples by MRM

Routine quantitation of unknown samples can be performed using a calibration curve that covers a much narrower concentration range of the endogenous analyte. We recommend defining a standard sample that is prepared alongside every batch of unknown samples. The tryptic digest of this standard sample is then used to generate the calibration curve sample concentrations. We routinely use calibration curves that span a 30-fold concentration range [11]. This concentration range can be extended if the variation in target protein concentration is expected to exceed 30-fold in the typical samples that will be analyzed. For routine quantitation we perform a single LC/MRM-MS analysis of each unknown sample with the analysis of a solvent blank between unknown samples.

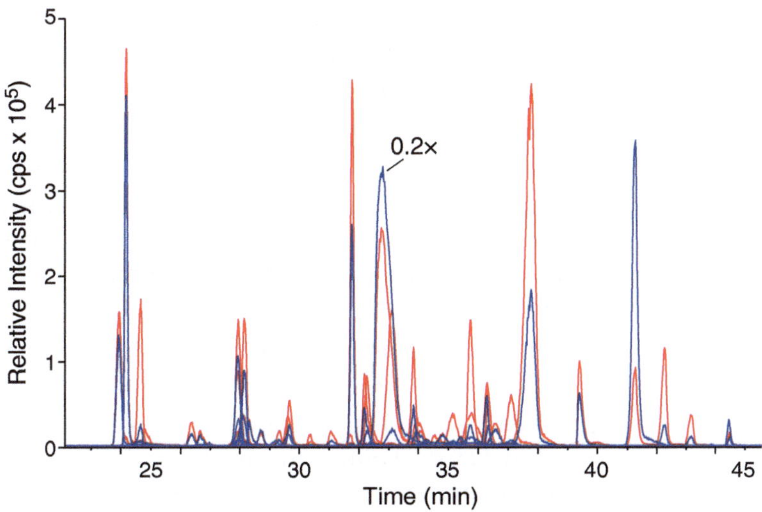

Fig. 6 XICs of all 45 MRM protein assays in a single 60 min LC/MRM-MS analysis of 1 μg (14.29 nL) of plasma tryptic digest spiked with a concentration-balanced mixture of 45 SIS peptide internal standards. MRM ion pair XICs of endogenous peptides are in *gray*, and SIS peptides are in *black*. Signal intensity of the endogenous albumin peptide has been rescaled by a factor of 0.2. Reprinted from ref. 11, with permission

MRM assays are very amenable to multiplexing. This workflow for the generation of characterized MRM assays can be developed for multiple proteins in parallel. An example of this approach is the multiplexed MRM quantitation of 45 human plasma proteins in a single 60 min LC/MRM-MS analysis (Fig. 6). In addition to multiplexing, additional effort can be spent on optimizing the chromatography conditions specific to your analyte. A 60 min analysis is a good starting point, but it is unnecessarily long when analyzing only a few peptides. We have reduced the analysis of our 45 peptide, multiplexed MRM assay to 30 min without sacrificing sensitivity. With the advent of microfluidics-based LC separations integrated with nanospray ionization, much shorter analysis times are possible (<10 min) when analyzing fewer peptides or lower complexity samples [28].

3.7 Sample Preparation for MRM Analysis

3.7.1 EDTA-Plasma Collection Sample

1. Collect blood by venous puncture using a 21-gauge needle.
2. Collect blood using the appropriate blood collection tubes.
3. Mix blood with the EDTA and protein stabilizers in the tubes.
4. Centrifuge at 1,000 × *g* for 15 min at 22 °C to pellet the cells and collect plasma.
5. Centrifuge again at 1,000 × *g* for 15 min at 22 °C to remove any remaining cells.
6. Divide plasma into 1.0 mL aliquots in 2 mL screwtop microtubes and freeze immediately at −80 °C for storage.
7. The total elapsed time from collection to storage should ideally be less than 1 h.

3.7.2 Tryptic Digestion of EDTA-Plasma

1. Quickly thaw a frozen aliquot of plasma at room temperature (22 °C). Vortex samples to help them both rapidly thaw and resuspend any insoluble material.
2. Centrifuge at 13,000 × *g* for 10 min at room temperature to remove any insoluble material (protein aggregates, lipids).
3. Avoiding any pellets or lipid layers in the plasma sample, carefully transfer 5 μL of plasma (~350 μg of total protein) to a 1.5 mL polypropylene tube containing 45 μL of 25 mM ammonium bicarbonate (*see* **Note 16**).
4. Denature the plasma sample by adding 50 μL of sodium deoxycholate (10 % w/v, 25 mM ammonium bicarbonate).
5. To obtain a 1/100 dilution of raw plasma during digestions, the final target volume for the digest is 500 μL. Add 291 μL of 25 mM ammonium bicarbonate to the denatured plasma sample (*see* **Notes 17** and **18**).
6. Reduce disulfide bonds with the addition of 43.4 μL of 50 mM TCEP (*see* **Note 19**) to provide a final concentration of 5 mM.
7. Incubate at 60 °C for 30 min.
8. Alkylate sulfhydryl groups by adding 48.3 μL of 100 mM iodoacetamide to provide a final concentration of 10 mM iodoacetamide (*see* **Note 20**).
9. Incubate at 37 °C for 30 min in darkness.
10. Add 17.5 μL of sequencing-grade, porcine-modified trypsin (0.4 mg/mL, 25 mM ammonium bicarbonate) to give a 50:1 substrate:enzyme ratio (w:w).
11. Ensure samples are thoroughly mixed by vortexing for 5 s.
12. Pulse centrifuge samples for 30 s at 13,000 × *g* at room temperature (22 °C).
13. Digest for 16 h at 37 °C.
14. If SIS peptides are being added to this sample, proceed to Subheading 3.7.3. Otherwise, stop digestion by acidifying the samples with 500 μL of 1 % v/v formic acid and proceed to Subheading 3.8.

3.7.3 Spiking of Plasma Tryptic Digests with SIS Peptides

Ideally, SIS peptides would be added as early as possible in the sample preparation workflow. However, we have found that addition of SIS peptides prior to trypsin digestion results in elevated peak area ratios (light:heavy) when compared to those obtained with post-digestion addition of SIS peptides [14]. The increase in the peak area ratio increase is not predictable and varies between peptides. SIS peptides are either being degraded or chemically modified over the course of the digestion. To avoid this source of variation, SIS peptide mixture should always be added to samples post-digestion, i.e., during digest acidification and prior to desalting and concentration by solid-phase extraction (SPE), so that any losses incurred during sample handling still equally affect both endogenous and heavy peptides. Post-digestion addition of standards also requires less of the SIS peptide mixture since only as much sample as is required for LC/MRM-MS analysis needs to be prepared.

When spiking samples with SIS peptides, it is worth determining how many LC/MRM-MS analyses will need to be conducted of each sample to save on SIS peptide consumption. We recommend preparing enough spiked sample for ~15 LC/MRM-MS analyses. This generates more sample than is required for the initial analysis, with enough remaining for later reanalysis. One microgram of total protein (equivalent to 14.3 nL of raw plasma) can be injected per LC/MRM-MS analysis, without compromising chromatographic performance, when using a 75 μm ID × 150 mm C_{18} reversed-phase analytical HPLC column.

1. The amount of SIS peptide to add to a plasma digest depends on the endogenous concentration of the target protein. A good starting point, and the example that we will initially use, is to aim for 100 fmol of SIS peptide to be injected on column per analysis. This equates to a SIS peptide concentration in plasma of 7 pmol/μL.
2. Dilute SIS peptides in 0.1 % v/v formic acid to a concentration of 100 fmol/μL.
3. Transfer 21.4 μL of plasma digest to a fresh 1.5 mL polypropylene tube. Since each digest contains a 1/100 dilution of plasma, this volume contains 0.214 μL of raw plasma which is enough for 15 sample injections at 14.29 nL per injection (*see* **Note 21**).
4. Store the remaining peptide solution and tryptic digest at −80 °C (*see* **Note 22**).
5. Add 15 μL of each 100 fmol/μL SIS peptide to the tube containing the 21.4 μL of sample.
6. Add an additional 100 μL of 1 % v/v formic acid to acidify the digest and precipitate sodium deoxycholate.

3.8 Solid-Phase Extraction of Tryptic Digests

Solid-phase extraction (SPE) using a reversed-phase resin is a necessary step to desalt and concentrate peptides prior to LC/MRM-MS analysis. Sodium deoxycholate is insoluble in acidic solutions, so sodium deoxycholate can be removed from acidified digests by centrifuging for 10 min at 13,000 × *g* at room temperature (22 °C).

1. Prepare one SPE cartridge for each sample by wetting the resin with 1 mL of methanol (*see* **Note 23**).
2. Apply a vacuum to draw the methanol through the resin.

3. Prior to sample application, equilibrate resin by drawing 1 mL of water through each cartridge.
4. Transfer the supernatant (exactly 100 μL) from the acidified sample (avoiding the deoxycholate pellet) to a 1.5 mL polypropylene tube and bring to 1 mL with 0.1 % FA. This dilutes the sample for more effective binding (*see* **Note 24**).
5. Transfer each acidified and diluted plasma digest sample to an SPE cartridge.
6. Apply a vacuum and draw the sample through the resin.
7. Wash each sample by drawing 1 mL of water through each cartridge.
8. Elute each sample into a clean polypropylene tube with 200 μL of 50 % v/v acetonitrile, 0.1 % v/v formic acid.
9. Freeze the samples at −80 °C and either lyophilize or vacuum concentrate the samples to dryness.
10. Store samples at −80 °C until analysis.

3.9 LC/MRM-MS Analysis of Tryptic Plasma Digests

These instructions require a triple quadrupole mass spectrometer equipped with a nanospray ionization source that is coupled to a nanoflow rate HPLC with an autosampler. For the following section, the acquisition parameters for an Applied Biosystems/MDS Sciex 4000 QTRAP controlled by Analyst 1.5 have been provided. The nano-HPLC should be configured so the autosampler is in line with the gradient nano-pump in order to permit direct injection of sample onto the analytical column. This maximizes chromatographic resolution and prevents loss of highly hydrophilic peptides.

3.9.1 Optimization of Mass Spectrometric Acquisition Parameters

1. Operation of a 4000 QTRAP in positive ion mode with an ion spray voltage of 1,800–2,000 V will usually establish stable spray at a 300 nL/min flow rate of 100 % solvent A (*see* **Note 25**).
2. CAD gas (collision gas) pressure should be tuned to obtain an MS operating pressure of 3.5×10^{-5} Torr during all MRM scans (*see* **Note 26**).
3. For specificity when analyzing a sample as complex as a plasma digest, MRM acquisition methods must be constructed using unit resolution (0.6–0.8 Da FWHH) for quadrupole 1 and quadrupole 3 (Q1 and Q3), respectively.
4. Additional 4000 QTRAP settings we find to be optimal for nanospray LC/MRM-MS analysis are 25 psi curtain gas, 150 °C interface heater temperature, and 1–3 psi sheath gas (GS1) (*see* **Note 27**).
5. Selection of MRM Q1/Q3 ion pairs for quantifying target peptides is outlined in detail in Subheading 3.7.

6. When constructing an MRM acquisition method use a dwell time of 20 ms for each MRM ion pair. Do not exceed a total cycle time of 2 s when monitoring multiple MRM ion pairs to ensure that each ion pair is sampled frequently enough to permit acquisition of 10–15 points across the elution profile of a peptide.
7. Several triple quadrupole manufacturers have recently incorporated the ability to enable and disable the monitoring of MRM ion pairs based on retention time constraints ("scheduled MRM"). This is beneficial when the list of MRM ion pairs is too long to permit a 2 s cycle time. If available, we recommend retention-based scheduling of MRM ion pairs using a target scan time of 2 s with a 4 min MRM detection window.

3.9.2 Online HPLC Parameters and Sample Analysis

1. For routine analysis of plasma digests, a 60 min LC/MRM-MS analysis is a good starting method that provides excellent resolution of a high-complexity sample such as trypsin-digested plasma.
2. Reconstitute samples in 11 μL of 0.1 % v/v formic acid, to obtain a final concentration of ~1 μg/μL, based on an initial plasma protein concentration of 70 mg/mL.
3. Inject a 1 μL volume of the sample onto the analytical column for 6 min at a flow rate of 300 nL/min, 100 % solvent A (*see* **Note 28**).
4. To separate the sample, use a 32-min linear gradient from 0 to 23 % solvent B, followed by a 9-min linear gradient from 23 to 43 % solvent B.
5. Strip the column by increasing solvent B from 43 to 80 % over 2 min. Hold solvent B at 80 % for 2 min, then decrease solvent B to 0 % over 2 min, and re-equilibrate the column at 100 % solvent A for 8 min prior to initiating the next sample pickup.
6. We recommend running a solvent blank between sample injections to both prevent and assess sample carryover.

3.10 MRM Data Analysis

Analysis of MRM data is quite simple compared to most other mass spectrometric analyses. The output of MRM analyses is an elution profile, or chromatogram of each MRM Q1/Q3 ion pair. Data analysis involves recognition and integration of the peak that is specific to the target analyte.

In MRM workflows that employ SIS peptides, the task of analyte peak recognition is quite straightforward since the retention time of the SIS peptide is known and the endogenous peptide must co-elute with its heavy counterpart. Quantitation software to accomplish this task is available from each vendor of triple quadrupole mass spectrometers. Once again, we present our recommendations using data output from an AB/MDS Sciex 4000 QTRAP.

1. Process the MRM data using MultiQuant 1.2 software (Applied Biosystems) using the MQL algorithm for peak integration.
2. Default peak integration settings we recommend are to detect the endogenous peak using a 2 min retention time window based on the SIS peptide's retention time.
3. Enable "report largest peak," use a peak-splitting factor of 2, and default values for noise percentage and baseline subtraction window.
4. It is strongly recommended that all integrated chromatograms be manually inspected to ensure that the software actually recognized and integrated the correct peak for each MRM ion pair.
5. When analyzing and quantifying data, work with ratios of endogenous peak area to SIS peptide peak area (light:heavy).
6. Linear regression analysis of calibration curves can be conducted using a weighting option of $1/x$ or $1/x^2$ (*see* **Note 29**).

4 Notes

1. Sigma-Aldrich also has an excellent webpage that echoes these suggestions at http://www.sigmaaldrich.com/life-science/custom-oligos/custom-peptides/learning-center/peptide-stability.html.
2. In our experience, oxidation of Trp is minor when compared to oxidation of Met, and can be prevented by minimizing exposure to ambient temperature.
3. We recommend that the C-terminal amino acid always be used as the location for stable isotope-labeled amino acids. The y-fragment ion series of tryptic peptides is the dominant fragment ion series in low-energy CID spectra [24]. Incorporating the heavy label in the C-terminal Arg or Lys shifts the mass of the entire y-ions series of the peptide CID fragments. This allows both the Q1 and Q3 masses to differ for the endogenous and heavy forms of a peptide, which maximizes the specificity of each MRM ion pair.
4. We use TentaGel R resin is used for peptide synthesis. Although Rapp-Polymere (http://www.rapp-polymere.com/) does not sell TentaGel R resin conjugated to isotopically labeled Arg or Lys amino acids they will couple custom amino acids to TentaGel R resin as a service.
5. Although double coupling is not required, we routinely double-couple amino acids to maximize coupling efficiency regardless of the predicted coupling difficulty.
6. Peptides that are difficult to resolubilize can sometimes be solubilized with higher acetonitrile concentrations (up to 15 %)

and probe sonication. However options are limited since the conditions used need to be compatible with RP-HPLC for purification and for later use in MRM workflows.

7. We use MALDI-TOF MS for this screening.

8. The Protein Prospector Web site has a useful web-based utility called *MS-Product* (http://prospector.ucsf.edu/prospector/cgi-bin/msform.cgi?form=msproduct) that can be used to calculate peptide fragment ions. Alternatively, an excellent Windows-based freeware utility called *Molecular Weight Calculator* contains this functionality and can be downloaded at this Web site: http://omics.pnl.gov/software/MWCalculator.php.

9. In general, low-mass ions (<300 Da) such as a_2 and b_2 ions, although they often give strong signals, should not be selected due to a higher probability of interference. Also be careful with y_2 ions. Of course if no other options are available, these can be used if they are first checked for interferences as described in the text.

10. By collecting data for these ion pairs, we can determine what percentage of signal can be attributed to isotopic impurity of the SIS peptide. In our experience this is less than 0.05 % of the SIS peptide signal when analyzing peptides containing stable isotope-labeled Arg and Lys (98 % isotopic enrichment) from Cambridge Isotope Labs.

11. If endogenous levels of the target protein are undetectable in your standard sample and a recombinant source of the protein is available, you can repeat this analysis by spiking the recombinant protein into plasma prior to trypsin digestion. To ensure detection of the endogenous peptides, spike enough protein to obtain 50–100 fmol of protein on column per analysis (3.5–7 pmol of protein per microliter of plasma).

12. Ideally this would be a y-ion with an m/z larger than the m/z of the precursor ion. This reduces the possibility of interfering signals.

13. This approach requires a protein to be naturally detectable in a sample and may not be suitable for low-abundance proteins.

14. The amount of digest required to generate the elevated concentrations of the endogenous analyte (samples 1 and 2 in Table 2) can be scaled down as long as the amount of SIS peptide used is proportionately reduced.

15. Using the example from Fig. 4, if the endogenous peptide in sample 3 (Table 2) has an average measured peak area ratio of 0.6, then we assume 100 % tryptic digestion efficiency and calculate the endogenous concentration of the peptide in the standard plasma sample to be 21 pmol/μL.

16. The accuracy of this pipetting step is critical for minimizing the technical variation introduced during sample preparation. Since

plasma is such a highly proteinaceous sample (60–80 mg/mL), use an accurately calibrated 10 μL pipette to perform this pipetting step. To minimize excess sample carryover on the outside of the pipette tip, quickly wipe the outside of the tip with a KimWipe, being careful not to wick the sample out of the pipette tip.

17. If you have a pipettor capable of reproducibly and accurately delivering exactly 290.8 μL, add this amount. If not, add 291 μL.
18. This additional dilution step lowers the protein concentration to prevent protein precipitation difficulties during reduction and alkylation.
19. This is a net 22.5-fold excess over the ~26 mM concentration of protein cysteine in plasma calculated from the known protein concentrations.
20. This is a net 50-fold excess over plasma protein cysteine.
21. For our standard LC/MRM-MS analysis using a 75 μm ID × 150 mm C_{18} analytical column, we inject a volume of sample that is equivalent to 14.29 nL of plasma (1/70 of 1 μL = 1 μg).
22. For prolonged storage of peptides we find it most practical to store them frozen in solution at −80 °C in 1.5 mL screwtop microtubes that contain O-rings to prevent sample loss due to sublimation. It is best to minimize the number of freeze–thaw cycles that the master stock is subjected to. As a result, we find it most practical to keep SIS peptide master stocks as ~500 nmol aliquots in 30 % v/v acetonitrile, 0.1 % v/v formic acid. Working stocks of SIS peptides (100 and 10 μM) are kept in at −80 °C in 0.1 % v/v formic acid for more routine use.
23. We find that Waters Oasis HLB SPE cartridges are excellent at retaining hydrophilic peptide species and are tolerant of air entering the resin bed making them compatible with standard vacuum manifolds.
24. If we use 100 μL here, considering the above instructions, the sample should be rehydrated in 11 μL of 0.1 % FA before MS analysis for a 1 μg on column with 100 fmol SIS peptide.
25. A standard red pen laser is essential for visualizing the Taylor cone at nL/min flow rates. Tune the IS (Ion Spray) voltage while monitoring the shape and stability of the Taylor cone to achieve a stable spray.
26. We find that lowering the operating pressure of the mass spectrometer to 3.5×10^{-5} from the typical pressure of 4.5–5.0×10^{-5} Torr that is typically recommended increases MRM signal intensity.
27. We find that both spray stability and life span of the uncoated fused silica emitter tips (20 μm inner diameter, 10 μm tip, New

Objective, Woburn, MA) can be dramatically improved by using addition of a post-column, makeup solvent (50 nL/min, 80/10/10 v/v/v isopropanol/methanol/water) using a PicoPlus 11 syringe pump (Harvard Apparatus, Holliston, MA).

28. For our standard LC/MRM-MS analysis to confirm peptide co-elution, we inject 14.29 nL of plasma per analysis (1/70 of 1 μL≈1 μg).
29. It is standard to use linear regression weighting which prevents higher values from over-contributing to the linear regression in a dilution series [11].

Acknowledgements

This work was funded by a platform grant from Genome Canada and Genome British Columbia.

References

1. Wei J, Sun J, Yu W, Jones A, Oeller P, Keller M, Woodnutt G, Short JM (2005) Global proteome discovery using an online three-dimensional LC-MS/MS. J Proteome Res 4:801–808
2. Kondrat RW, McClusky GA, Cooks RG (1978) Multiple reaction monitoring in mass spectrometry/mass spectrometry for direct analysis of complex mixtures. Anal Chem 50:2017–2021
3. Anderson NL, Anderson NG (2002) The human plasma proteome: history, character, and diagnostic prospects. Mol Cell Proteomics 1:845–867
4. Anderson L (2005) Candidate-based proteomics in the search for biomarkers of cardiovascular disease. J Physiol 563:23–60
5. Ackermann BL, Berna MJ (2007) Coupling immunoaffinity techniques with MS for quantitative analysis of low-abundance protein biomarkers. Expert Rev Proteomics 4:175–186
6. Kamiie J, Ohtsuki S, Iwase R, Ohmine K, Katsukura Y, Yanai K, Sekine Y, Uchida Y, Ito S, Terasaki T (2008) Quantitative atlas of membrane transporter proteins: development and application of a highly sensitive simultaneous LC/MS/MS method combined with novel in-silico peptide selection criteria. Pharm Res 25:1469–1483
7. Barnidge DR, Goodmanson MK, Klee GG, Muddiman DC (2004) Absolute quantification of the model biomarker prostate-specific antigen in serum by LC-MS/MS using protein cleavage and isotope dilution mass spectrometry. J Proteome Res 3:644–652
8. Kuhn E, Wu J, Karl J, Liao H, Zolg W, Guild B (2004) Quantification of C-reactive protein in the serum of patients with rheumatoid arthritis using multiple reaction monitoring mass spectrometry and 13C-labeled peptide standards. Proteomics 4:1175–1186
9. Zhang F, Bartels MJ, Stott WT (2004) Quantitation of human glutathione S-transferases in complex matrices by liquid chromatography/tandem mass spectrometry with signature peptides. Rapid Commun Mass Spectrom 18:491–498
10. Kirkpatrick DS, Gerber SA, Gygi SP (2005) The absolute quantification strategy: a general procedure for the quantification of proteins and post-translational modifications. Methods 35:265–273
11. Kuzyk M, Smith D, Yang J, Cross T, Jackson A, Hardie D, Anderson L, Borchers C (2009) MRM-based, multiplexed, absolute quantitation of 45 proteins in human plasma. Mol Cell Proteomics 8:1860–1877
12. Walsh GM, Lin S, Evans DM, Khosrovi-Eghbal A, Beavis RC, Kast J (2009) Implementation of a data repository-driven approach for targeted proteomics experiments by multiple reaction monitoring. J Proteomics 72:838–852
13. Mallick P, Schirle M, Chen SS, Flory MR, Lee H, Martin D, Ranish J, Raught B, Schmitt R, Werner T, Kuster B, Aebersold R (2007) Computational prediction of proteotypic peptides for quantitative proteomics. Nat Biotechnol 12:125–131

14. Skyline_SRM/MRM_Builder (2009) https://brendanx-uw1.gs.washington.edu/labkey/wiki/home/software/Skyline/page.view?name=default. Accessed 15 Aug 2009
15. MacLean B, Tomazela D, Finney G, Chambers M, Shulman N, Prakash A, Peterman S, Maccoss MJ (2009) Automated creation and refinement of complex scheduled SRM methods for targeted proteomics. Presented at the 57th ASMS conference on mass spectrometry and allied topics, Philadelphia, PA, 31 May to 4 June 2009
16. Prakash A, Tomazela DM, Frewen B, MacLean B, Merrihew G, Peterman S, MacCoss* MJ (2009) Expediting the development of targeted SRM assays: using data from shotgun proteomics to automate method development. J Proteome Res 8:2733–2739
17. MRMpilot. https://products.appliedbiosystems.com/ab/en/US/adirect/ab?cmd=catNavigate2&catID=605354&tab=DetailInfo. Accessed 15 Aug 2009
18. Sherwood C, Eastham A, Peterson A, Eng JK, Shteynberg D, Mendoza L, Deutsch E, Risler J, Lee LW, Tasman N, Aebersold RL, Lam H, Martin DB (2009) MaRiMba: a software application for spectral library-based MRM transition list assembly. J Proteome Res 8:4396–4405
19. Mead JA, Bianco L, Ottone V, Barton C, Kay RG, Lilley KS, Bond NJ, Bessant C (2009) MRMaid, the web-based tool for designing multiple reaction monitoring (MRM) transitions. Mol Cell Proteomics 8:696–705
20. Martin DB, Holzman T, May D, Peterson A, Eastham A, Eng J, McIntosh M (2009) MRMer, an interactive open source and cross-platform system for data extraction and visualization of multiple reaction monitoring experiments. Mol Cell Proteomics 7:2270–2278
21. Xu Y, Lazar IM (2009) MRM screening/biomarker discovery with linear ion trap MS: a library of human cancer-specific peptides. BMC Cancer 9:96
22. Picotti P, Lam H, Campbell D, Deutsch EW, Mirzaei H, Ranish J, Domon B, Aebersold R (2008) A database of mass spectrometric assays for the yeast proteome. Nat Methods 5:913–914
23. MIDAS_workflow_designer. https://products.appliedbiosystems.com/ab/en/US/adirect/ab?cmd=catNavigate2&catID=604874&tab=DetailInfo. Accessed 15 Aug 2009
24. Janecki DJ, Bemis KG, Tegeler TJ, Sanghani PC, Zhai L, Hurley TD, Bosron WF, Wang M (2007) A multiple reaction monitoring method for absolute quantification of the human liver alcohol dehydrogenase ADH1C1 isoenzyme. Anal Biochem 369:18–26
25. Green JM (1996) A practical guide to analytical method validation. Anal Chem 68: 305A–309A
26. Lin S, Shaler TA, Becker CH (2006) Quantification of intermediate-abundance proteins in serum by multiple reaction monitoring mass spectrometry in a single-quadrupole ion trap. Anal Chem 78:5762–5767
27. Keshishian H, Addona T, Burgess M, Kuhn E, Carr SA (2007) Quantitative, multiplexed assays for low abundance proteins in plasma by targeted mass spectrometry and stable isotope dilution. Mol Cell Proteomics 6:2212–2229
28. Yin H, Killeen K (2007) The fundamental aspects and applications of Agilent HPLC-Chip. J Sep Sci 30:1427–1434
29. Roepstorff P, Fohlman J (1984) Proposal for a common nomenclature for sequence ions in mass spectra of peptides. Biomed Mass Spectrom 11:601

Chapter 5

Mass Spectrometric Profiling of Low-Molecular-Weight Proteins

Matthias Rainer, Constantin Sajdik, and Günther K. Bonn

Abstract

Tracing potential biomarkers through proteomics has been further developed and is nearing realization. The whole sequence of human proteome is becoming better understood with the passage of time. However, it is a long way to go to pinpoint biomarker proteins out of complex biofluids and use them for clinical diagnosis, prognosis, and therapeutic applications. From that point of view, the high hopes put in proteomics have not been fulfilled yet. The key reasons for that is the complexity of the proteome and the limited technologies in terms of specificity and reproducibility. Thus, major focus is put on the development of novel innovative analytical techniques in the field of life science, using high-performance single- and multidimensional separation and enrichment methods, such as solid-phase extraction (SPE), liquid chromatography (HPLC), or capillary electrophoresis (CE) coupled to mass spectrometry (MS). A newly emerged technology, termed as material-enhanced laser desorption/ionization (MELDI) meets basic requirements and is applied to reduce the complexity of proteomic samples while liquid chromatography (LC) is used for separation and fractionation, followed by identification with MS/MS including database searching analysis. Different MELDI carriers are employed as support materials to specifically bind peptides and proteins from biofluids like serum or urine. The MELDI approach supports automated routine analysis by means of liquid handling robots for high-throughput applications leading to higher reproducibility, crucial for a successful identification of disease markers with MALDI-TOF MS. Such promising new methods and further technical developments will be necessary to answer the high-wrought expectations on the field of proteomics.

Key words MELDI, MALDI, SELDI, Proteomics, Biofluids, Biomarker, Mass spectrometry

1 Introduction

One of the major target applications of proteomics is to evaluate disease-related biomarkers, which by over- or underexpression of certain peptides or proteins help to distinguish between healthy and diseased samples for early diagnosis and which even allow the determination of the advancement of the case. Biomarkers are often low-molecular-weight proteins secreted into the bloodstream as a result of the disease process [1]. In addition, biomarkers should allow an early diagnosis as well as assumptions on stage/progression,

Helena Bäckvall and Janne Lehtiö (eds.), *The Low Molecular Weight Proteome: Methods and Protocols*, Methods in Molecular Biology, vol. 1023, DOI 10.1007/978-1-4614-7209-4_5,

severity, and prognosis of a disease. Traditionally, scientists have searched for disease biomarkers using a one-at-a-time approach, by focusing only on one particular marker. However, due to the complexity of many diseases, it is very likely that the use of multiple biomarkers in screening and diagnosis will be necessary to produce unequivocal results [2]. Hence, biomarker research is shifting to so-called profiling methods, allowing the simultaneous measurement of a range of markers. It is suspected that such marker patterns will allow a statistically more stringent differentiation and a better classification of patient groups. This should lead to an improvement of early disease detection by reducing the number of false positive or negative results. Biomarkers are often low-molecular-weight proteins, shed into the bloodstream by mutated, apoptotic, or necrotic cells from disease-affected organs. So it seems convenient to use plasma/serum as a diagnostic specimen, also due to its easy and inexpensive purchase. Unfortunately, it is technically difficult to detect the small and less abundant proteins: firstly, because of the generally enormous dynamic range of protein concentrations in blood (estimated more than nine orders of magnitude) and secondly, due to the fact that only 22 of the most abundant proteins make up 99 % of the whole serum proteome, while the more specific low-abundant disease-related proteins are remaining in that 1 %. To skirt those obstacles, new methods have been developed. Those alternative approaches are aiming to remove the high-abundant proteins prior to analysis (especially prior MS). However, the reproducibility and effectiveness of many depletion methods have always been questioned and there are serious concerns about a loss of important markers stuck to the high-abundant carrier proteins. Albumin for example has a strong ability to bind high- as well as low-abundant proteins and peptides [3]. Moreover extensive sample handling during the depletion process increases the chance of sample loss, protein degradation, and modification artifacts causing significant sample-to-sample variations. In consideration of the complexity of serum, other body fluids such as urine, cerebrospinal fluid, saliva, bronchoalveolar lavage fluid, synovial fluid, tear fluid, nipple aspirate fluid, and ammoniac amniotic fluid might be valuable alternative sources for biomarker discovery. Nevertheless there are a lot of bioanalytical approaches with a high potential to discover new relevant disease markers from biological samples. There are quite a lot of bioanalytical approaches which might have a very high potential for discovering relevant disease markers from biological samples. Especially proteomic pattern analysis can be a first step for the detection of novel diagnostic markers out of body fluids when comparing mass fingerprints from patients with those from control. Protein profiling refers to fast and high-throughput mass spectrometric analysis (screening) of biological samples containing intact proteins. Screening enables the analysis of a high number of samples in a short period of time and can thus be employed for mining out differences (and consequently potential biomarkers) between a huge number of diseased and healthy samples [4]. Surface-enhanced laser desorption/ionization (SELDI), which has originally been first introduced by Hutchens and Yip [5], has proven to be a useful screening tool for prepurification, selective adsorption, as well as pre-concentration of biological samples prior to MS evaluation. The complex protein mixture is subjected to equilibrated, derivatized mass spectrometric surfaces, exhibiting different chromatographic properties in terms of surface chemistry [6, 7]. According to the chemical nature of the target surface (hydrophilic, hydrophobic, ion-exchange, IMAC), sample constituents get selectively adsorbed upon incubation. After the removal of unspecifically bound proteins and a quick washing step for sample desalting, an appropriate matrix solution is added and the loaded target subjected to MALDI-MS analysis in order to obtain the respective protein profiles. The differences in protein profiles of disease- and control-related samples are caused by overexpression, abnormally shed proteins or protein fragments, modified proteins, proteolytically cleaved proteins, or degradation due to the proteosome pathway [8]. In 2006, a new form of laser desorption/ionization technique was introduced by Bonn et al. which is referred to as material-enhanced laser desorption/ionization (MELDI) [9, 10]. In MELDI, proteins get adsorbed onto the surface of a substrate material but, in contrast to SELDI, the physical and chemical properties of the surface material, including the particle morphologies and porosities, come into play. The substrate materials (MELDI materials) can be selected to provide a particular functionality, in order to attract a certain type of compound (proteins and peptides) out of complex biological samples, which can be then profiled by MALDI-MS. The MELDI material gets first equilibrated and incubated with biological samples (e.g., serum, urine). After washing of unbound components, the protein-loaded MELDI material gets directly subjected to MALDI-MS analysis (pattern analysis). In a parallel approach the enriched proteins can be eluted, enzymatically digested, and separated by RP-HPLC for label-free quantification and MS/MS identification. The ability of MELDI to distinguish prostate cancer from control was proven by a comprehensive biomarker study based on prostate cancer. Prostate cancer is the most common malignancy in men in Western countries [11], and early diagnosis is still based on the serum test for prostate-specific antigen (PSA), a test with limited disease specificity (35 %) [12, 13]. Therefore, the search for new and more reliable biomarkers to stratify disease onset and progression remains a challenge. It is very likely that multiple biomarkers will be required to improve early detection, diagnosis, and prognosis. This chapter will provide a general introduction about the MELDI approach.

2 Materials

For a MELDI-based biomarker study on prostate cancer, spherical cellulose beads (8 μm in diameter) with IMAC-Cu^{2+} functionalities were employed to specifically bind serum constituents [14]. The MELDI kits were purchased from PhyNexus Inc (San Jose, CA) and contain permeable pipette tips (PhyTips) which are filled with cellulose IMAC-Cu^{2+} resin.

2.1 MELDI Sample Preparation

1. Phosphate-buffered saline (PBS): 0.01 M Na_2HPO_4/NaH_2PO_4 and 0.15 M NaCl (pH 7.4), (PhyNexus Inc, San Jose, USA).
2. Denaturing solution J: 8 M urea containing 1 % 3-[(3-cholamidopropyl)dimethyl-ammonio]-1-propanesulfonate (CHAPS) in PBS (PhyNexus Inc, San Jose, USA).
3. Denaturing Solution K: 1 M urea containing 0.125 % CHAPS in PBS (PhyNexus Inc, San Jose, USA).
4. Washing solution L: PBS (PhyNexus Inc, San Jose, USA).
5. Washing solution B: Deionized water.
6. MALDI matrix is prepared by dissolving 60 mg of sinapinic acid in 3 ml of 50 % ACN and 0.1 % TFA.
7. Angiotensin I (1 μg/μl) and cytochrome C (1 μg/μl).
8. Human blood serum samples (*see* **Notes 1–4**).

2.2 MALDI-TOF MS

1. MALDI target, MTP 384 ground steel TF (Bruker Daltonics, Bremen, Germany).
2. For external MS calibration the following proteins are used (5 pmol/μl of each protein): insulin (5,734.51), ubiquitin I (8,565.76), cytochrome C (12,360.97), and myoglobin (16,952.30).
3. The protein-loaded cellulose resin is directly analyzed by MALDI-TOF MS (Ultraflex I, Bruker Daltonics, Bremen, Germany) equipped with a 337 nm nitrogen laser (50 Hz). All serum samples are measured in linear mode and the detector energy is set to 1,600 V. Data are collected by averaging 500 laser shots and analyzing mass region from 2 to 20 kDa. The validation of all data obtained, including baseline subtraction of the TOF data and external calibration using the protein standard, and all further data processing are carried out by using Flex analysis 2.4 post-analysis software and for data acquisition by Flex control 2.4 (Bruker Daltonics, Bremen, Germany).

2.3 Tryptic Digest on the Cellulose Resin

1. Urea (≥99.5 %), dithiothreitol (DDT, ≥99.0), iodoacetamide (≥98.0 %), trifluoroacetic acid (TFA, for protein sequence analysis), and *n*-octylglucopyranoside nOGP (≥99.0 %).
2. Trypsin sequencing grade modified (Promega Biosciences, San Luis Obistpo, CA, USA).

2.4 LC-ESI-MS Analysis of Digested Fractions

1. For separation: Monolithic poly[*p*-methylstyrene-co-1,2-bis (*p*-vinylphenyl)ethane] capillary column (80 mm × 0.2 mm).
2. Acetonitrile HPLC grade (Merck Darmstadt, Germany).
3. Deionized water Barnstead NANOpure Infinity Water Purification System (Barnstead, Boston, MA, USA).
4. Dionex Ultimate 2D-μHPLC system. Eluates are separated using RP conditions (solvent A, 0.1 % TFA in water; solvent B, 0.1 % TFA in ACN).
5. Quadrupole ion trap mass spectrometer (LTQ, Thermo Finnigan, San Jose, CA, USA) equipped with a nano-electrospray ion source.

3 Methods

3.1 Automation of MELDI Sample Preparation

All steps of sample preparation are completely performed on a liquid handling robotic system to achieve highest reproducibility (*see* **Note 5**). For that a PhyNexus MEA™ Personal Purification System (San Jose, USA) is adapted. The robotic system allows a fully programmable positioning of the 12-channel pipette and the ability to place sample, extraction tips, washing solutions, and elution plates in any available position. Up to 48 biological samples can be accommodated by the MEA system in one batch. Three separate methods (process A, B, and C) are required for performing MELDI sample preparation. These methods are designed to prepare biological samples for subsequent capture onto PhyTip MELDI columns (Fig. 1) followed by spotting the resin on a MALDI target. The following methods will prepare the sample for subsequent PhyTip MELDI column capture by adding denaturant, zwitterions, and buffer to the samples.

3.1.1 Process A: Preparation of Serum Samples

1. The MEA instrument is loaded with three boxes of transfer tips at positions 1–3 (Fig. 2).
2. 30 μl of each serum samples are placed into a well plate at position 8—chiller (*see* **Note 6**).
3. Denaturating solution J is placed in row A, solution K is added into row B, and buffer L is transferred to row C of a deep well plate at position 5.
4. Using transfer tips from position 1, 5 μl of denaturating solution J are added to each serum sample at position 8 and mixed using three cycles of intake/expel at a flow rate of 0.5 ml/min. Serum samples are continuously chilled at position 8. Tips are replaced in position 1.
5. 10 μl of sample preparation solution K are added to each serum sample followed by taking fresh transfer tips from position 2. Mixing the sample is performed in the same manner as before.

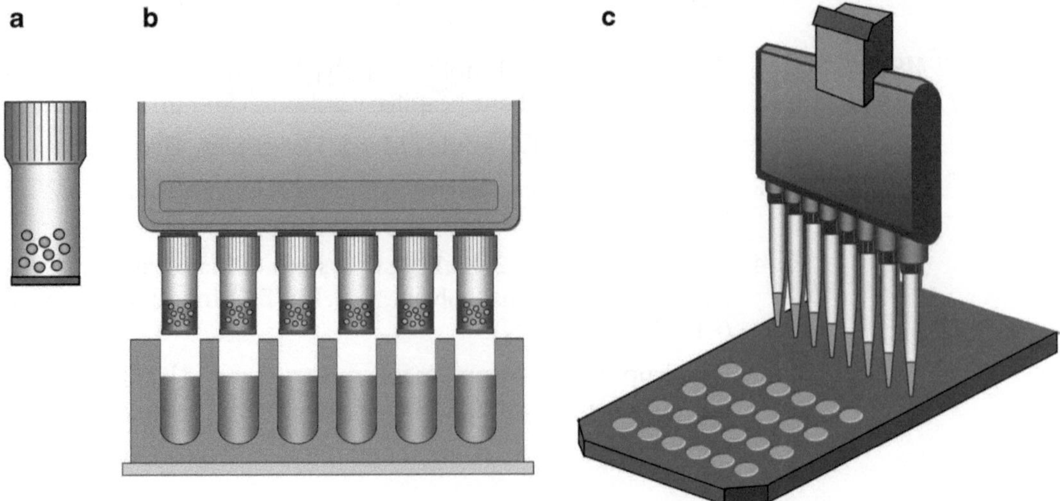

Fig. 1 MELDI sample preparation for IMAC-Cu^{2+} cellulose by employing a liquid handling robotic system. Equilibration of PhyTips in a deep well plate (**a**). PhyTip which is filled with IMAC-Cu^{2+} cellulose resin. A permeable membrane at the bottom of the column allows buffers to penetrate through the resin (**b**). After MELDI sample preparation, the protein-loaded resin is directly spotted on a ground steel target for further analysis by MALDI-MS (**c**)

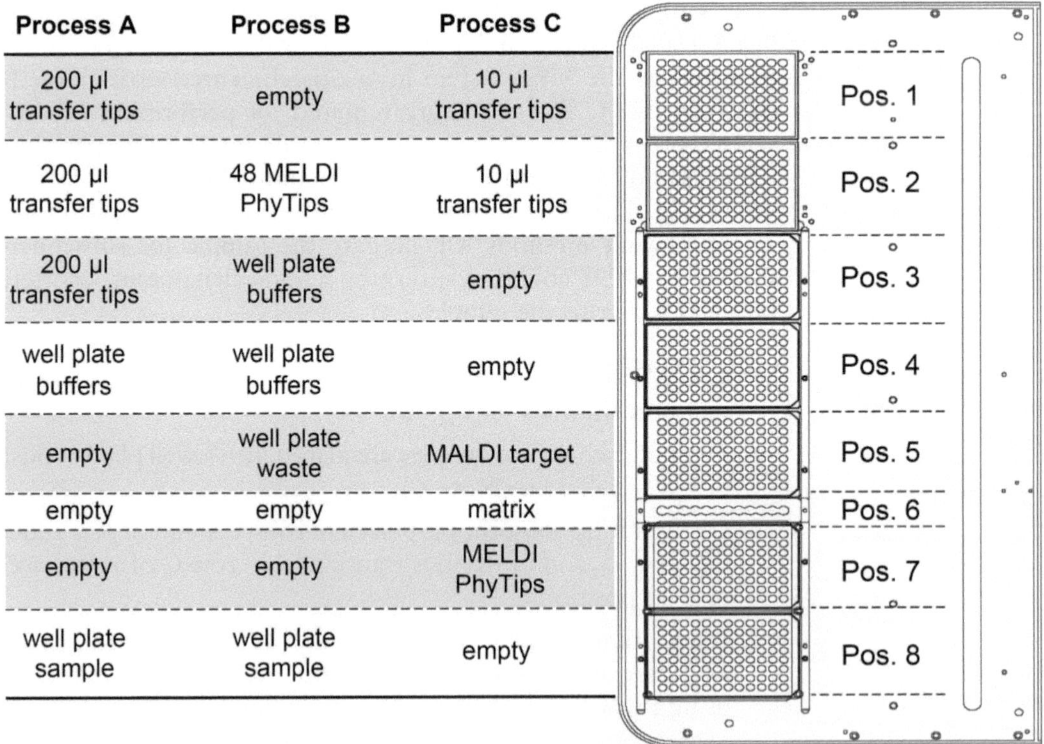

Process A	Process B	Process C	
200 µl transfer tips	empty	10 µl transfer tips	Pos. 1
200 µl transfer tips	48 MELDI PhyTips	10 µl transfer tips	Pos. 2
200 µl transfer tips	well plate buffers	empty	Pos. 3
well plate buffers	well plate buffers	empty	Pos. 4
empty	well plate waste	MALDI target	Pos. 5
empty	empty	matrix	Pos. 6
empty	empty	MELDI PhyTips	Pos. 7
well plate sample	well plate sample	empty	Pos. 8

Fig. 2 MELDI sample preparation for IMAC-Cu^{2+} cellulose PhyTips by employing a MEA liquid handling robotic system. The MEA system has eight different positions, whereas positions 7 and 8 are equipped with a chiller. For MELDI sample preparation three processes are required which can be classified in serum sample preparation, serum incubation, and on-target spotting of the cellulose resin

6. Transfer tips from position 3 are employed to add 80 μl of sample preparation solution L to the diluted serum samples at position 8 followed by a quick mixing step.

3.1.2 Process B: Enrichment and Purification

After the MEA system has added the required buffers to the serum samples, the system needs to be set up for affinity capture of the serum samples using PhyTip MELDI columns.

1. For that the MEA instrument is loaded with a box of PhyTip MELDI columns containing IMAC-Cu^{2+}-modified cellulose resin at position 2, a deep well plate containing washing solution L at position 3, and a deep well plate containing deionized water at position 4.
2. A row of MELDI PhyTips is picked up and washed with deionized water for five cycles by each 100 μl intake/expel at a flow rate of 1 ml/min at position 4.
3. The washed columns are then equilibrated at position 3 with PBS buffer in the same way as described before.
4. After equilibration, the MELDI columns are brought to position 8, where incubation of the prepared serum samples starts for ten cycles by 100 μl intake/expel at a flow rate of 250 μl/min.
5. To remove unbound serum constituents, MELDI columns are first washed for five cycles by each 100 μl intake/expel at 250 μl/min with PBS at positions 3 followed by desalting steps with deionized water at position 4. After each washing step the excess liquid is released as waste into a deep well plate at position 5 (*see* **Note 7**).
6. In a next step the moistened PhyTip columns are placed into a rack at position 7 (chiller) where they are kept cool and moist for process C.

3.1.3 Process C: Spotting of Cellulose Resin

For process C, the MEA deck is set up as follows:

1. A stainless steel target (MTP 384 ground steel) is placed at position 5 and MALDI matrix (sinapinic acid) is added into a row of PCR tubes at position 6. Matrix is spiked with 3 μl of angiotensin I (1 μg/μl) and 7 μl of cytochrome C (1 μg/μl) as internal standards for a later alignment process.
2. After the proteins have been captured by the MELDI PhyTips (process B), the MEA needs to be set up for spotting the resin onto MALDI targets. For that, the 200 μl pipettor needs to be exchanged by the 20 μl pipettor.
3. Using a row of transfer tips from position 1, 60 μl of matrix solution from position 6 are added to the moistened MELDI columns at position 7.

4. A suspension of cellulose and matrix is formed by cycling the mixture six times with 10 μl intake/expel at a flow rate of 1 ml/min.
5. Additionally a new row of transfer tips is placed into the resin and a 2 μl aliquot of slurry is soaked up into the tips.
6. The row of transfer tips containing the resin–matrix slurry is moved to position 5, directly above the MALDI target. Finally, the tips are lowered and 2 μl of resin/matrix is spotted onto the target (*see* **Note 8**).
7. The target is air-dried before MALDI-TOF MS analysis (*see* **Notes 9** and **10**).

3.2 Analysis of Human Sera from Prostate Cancer Patients

Serum protein profiling using MELDI is carried out to compare different mass fingerprints from individuals with benign and malignant prostate cancer. Spherical cellulose beads are employed as MELDI carrier materials for the specific binding of serum peptides and proteins, followed by their subsequent analysis with MALDI-TOF MS [14].

1. A training set was created containing 48 serum samples from benign and 48 serum samples from men with histologically confirmed prostate cancer. Every serum sample was measured in quadruplicate by the described MELDI approach (*see* **Note 11**). Sample preparation was carried out using the MEA™ Personal and Purification System.
2. For data interpretation including raw data pretreatment, normalization of spectra, internal signal alignment using prominent internal signal peaks, and peak-picking procedure the ClinProTools™ software (Bruker Daltonics, Germany) was used (Fig. 3). A k-nearest neighbor genetic algorithm was applied to identify statistically significant differences in protein peaks in the groups analyzed. After each model was generated, a 20 % leave-one-out cross-validation process was performed within the software (cross-validation refers to the accuracy of the software to correctly assign a random sample to the correct group). Cross-validated values were used for determining sensitivity and specificity of the classifications and yielded a sensitivity of 98 % for correct classification of prostate cancer, and specificity of 99 % for correct classification of benign serum samples.

3.3 A LC-MS-Based MELDI Workflow

The introduction of new bioanalytical methods including the mass spectrometric detection of biomarkers holds promise of providing diagnostic and prognostic information for cancer and other disease-related biological fluids. In this particular field of research a strong focus is laid on proteomic techniques that can be used to differentiate protein expression patterns between normal and different stages of disease. The MELDI approach is

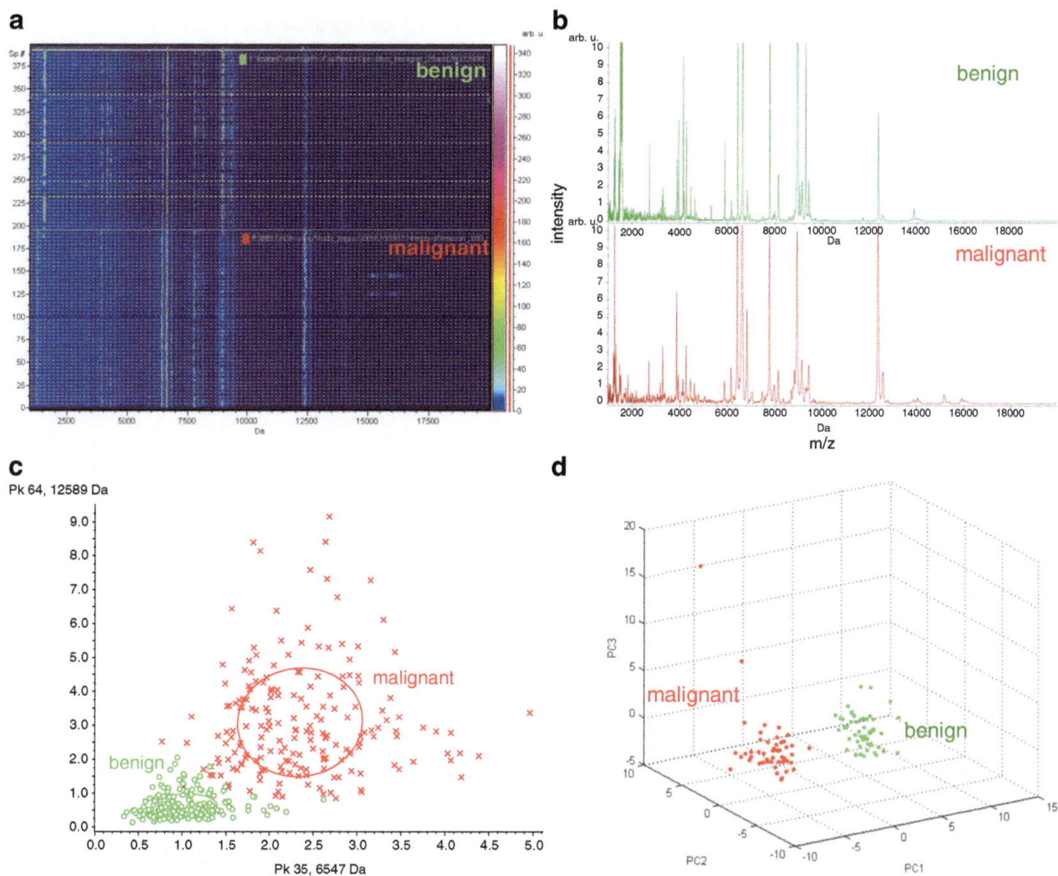

Fig. 3 Pseudo-gel view of 48 benign and 48 malignant serum profiles (**a**). All samples were measured in quadruplicate (*m/z* values along the *x*-axis and relative intensity along the *y*-axis). Average mass spectra of the corresponding data set (**b**). Differences can be already observed visually (*green,* benign; *red,* malignant). 2D peak distribution of the same set (**c**). Graphical representation of the classification, made on the basis of the intensity of the major differential signals on external datasets: *Circles* represent the predictions for the different regions. 3D view of Principal Component Analysis (PCA) scores plot analyzed by ClinProtools™ (**d**). *Green* spots represent benign spectra and the *red* spots represent malignant profiles

applied to reduce the complexity of proteomic samples while HPLC is used for fractionation, followed by enzymatic digestion of the enriched fractions before LC-ESI-MS identification. Bound proteins are eluted off the MELDI beads by solvent-dependent extraction for further mass spectrometric identification (Fig. 4). Proteins of molecular mass >4,000 have to be digested for analysis by μ-HPLC MS/MS with electrospray ionization or following off-line fractionation by MALDI-MS/MS.

1. For that, 1.5 mg of the protein-loaded IMAC-Cu^{2+} resin is suspended in 15 μl of 1 M NH_4HCO_3 solution followed by adding 3 μl of 40 mM *n*-octylglucopyranoside (nOGP) and 5 μl of 45 mM dithiothreitol (DTT).

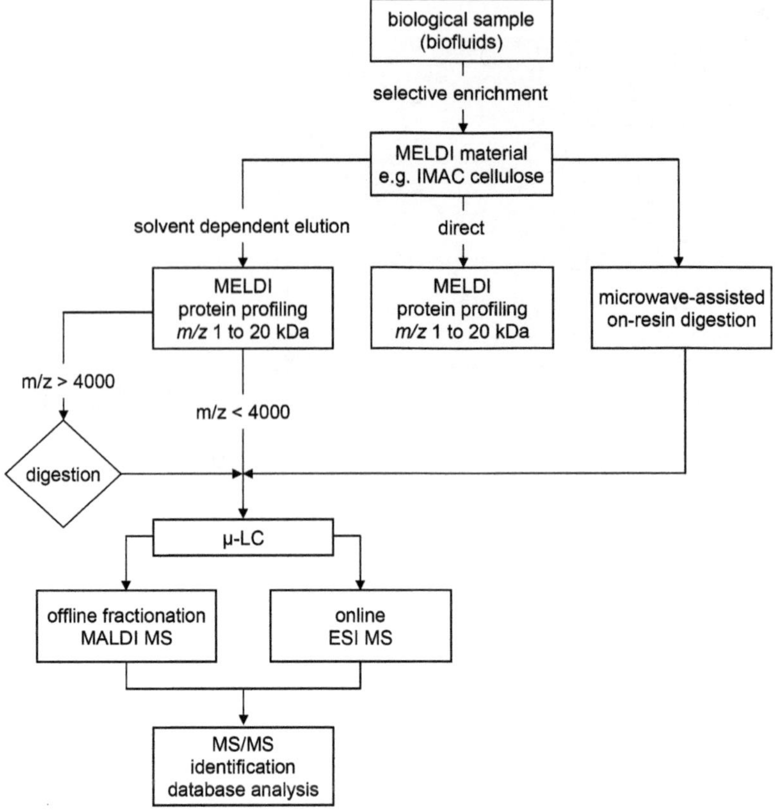

Fig. 4 MELDI-based workflow. In a first step both the MELDI carrier and the biological sample are incubated on a platform shaker. The MELDI carrier is directly spotted onto a MALDI target, followed by the addition of matrix prior to MALDI-MS measurement. Direct MALDI-MS analysis of the protein-loaded resin results in different fingerprints. Alternatively the bound proteins can be eluted off the MELDI material by solvent-dependent elution for further MALDI or LC-MS measurements. Proteins >4 kDa need to be digested for further MS/MS analysis. On the other hand an enzymatic on-resin digestion can be performed. The digested proteins are subjected to μ-LC separation followed by online ESI-MS or offline MALDI-MS

2. The suspension is incubated on a thermomixer (Eppendorf) at 58 °C for 15 min.
3. After denaturation, the suspension is cooled down to room temperature continued by the addition of 8 μl of 100 mM iodoacetamide.
4. Subsequently the suspension is placed into a dark chamber for 15 min at room temperature.
5. 5 μl of a trypsin solution (0.1 μg/μl) are added to the denatured and modified cellulose–protein complex.

6. Tryptic digestion is performed in a microwave at 100 W for 5 min.
7. The tryptic digest is stopped by adding 20 μl of a 1 % TFA solution (pH < 3.0).
8. The digested fraction is separated from the cellulose resin by 20 min centrifugation at 16,100 × *g*.
9. Finally, the digested fraction is subjected to a μ-HPLC system. For separation, monolithic capillary columns based on poly[*p*-methylstyrene-co-1,2-bis(*p*-vinylphenyl)ethane] are preferably employed which represent a serious alternative toward its silica-based counterpart [15] (*see* **Note 12**).
10. The analytical column is directly connected to the spray capillary (fused-silica, 90 μm OD, 20 μm ID; Polymicro Technologies) using a microtight union. Total ion chromatograms and mass spectra are recorded on a personal computer equipped with the LTQ Navigator software version 1.2 (Finnigan).
11. After mass spectrometry analysis one has to do a fast and reliable data processing. This is done by comparing MS-data with database-based data. The on-resin digested serum samples are label-free quantified and identified via μLC-MS followed by statistical data evaluation, delivering information about over- and under-expressed peptides and proteins (Fig. 5). Therefore, the Progenesis LC-MS software (Nonlinear dynamics, Newcastle, UK) is applied, creating a combined peptide map from all samples to detect regulated signals and identify the corresponding proteins by database search.

The ongoing development of MELDI for the analysis of clinically relevant samples represents an excellent tool for protein profiling and allows the identification and quantitation of potential markers by the additional use of LC-ESI-MS systems.

4 Notes

1. Do not add proteinase inhibitors after blood sampling as they might suppress the generation of disease markers by disease-specific proteinases.
2. Always use the same type of blood-collection tubes for serum collection.
3. To decrease the biological variance, a well-organized working plan including critical evaluation of sample quality, such as a complete history record, sampling condition, sample transportation, pretreatment, and storage, is needed.
4. No prior albumin or immunoglobulin depletion should be carried out as potential disease markers might be associated with albumin or other high-abundant serum proteins.

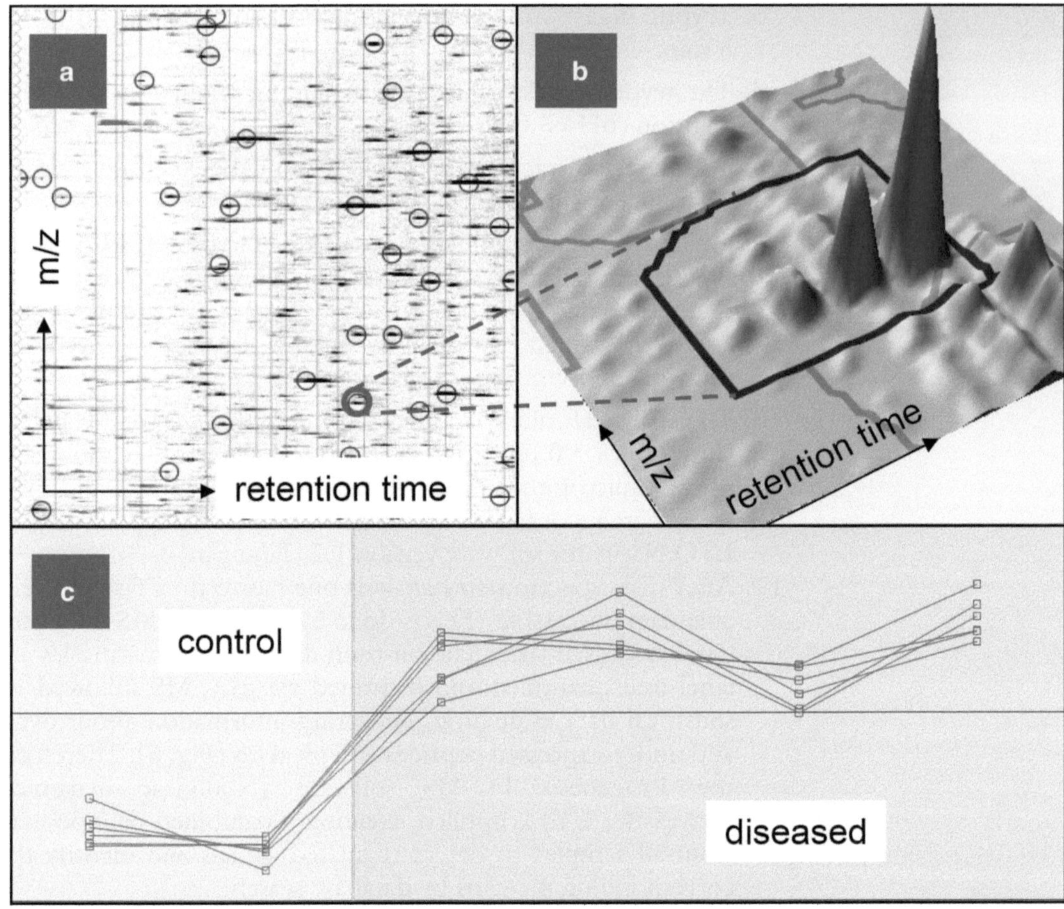

Fig. 5 Peptide map of two overlaid LC-MS runs (**a**). The virtual spots are located at *m/z* as monoisotopic molecular weights and at HPLC retention time as the gravity center of ion intensity (**b**). The intensity of the spots corresponds to the peak volume, defined as the integral of ion intensity. Up- and down-regulations of proteins can be evaluated in the different LC-MS runs for diseased and control samples (**c**)

5. A complete automation process of sample preparation is of utmost importance to minimize the analytical variance and to increase the robustness of the method.
6. Temperature strongly influences sample preparation. It is recommended to perform sample preparation at 4 °C.
7. To remove unspecific serum components, extensive washing steps have to be performed.
8. Prepare MALDI matrix always freshly.
9. It is recommended to quickly start MALDI profiling after performing MELDI sample preparation.
10. MALDI measurement should be carried out in automatic mode using a standardized method.

11. Never prepare only a single cohort of individual samples (prepare control and diseased samples in each run).
12. Due to the steadily growing interest in the field of proteomics, the new monolithic column supports can be an attractive tool for the fractionation complex samples, as they combine high chromatographic efficiency and short analysis time.

Acknowledgments

This work was supported by the Austrian Genome Program (Gen-AU), Vienna, Austria, and by the Project of the Austrian-Chinese Biomarker Platform, Ministry of Science and Research, Vienna. The authors would like to thank PhyNexus Inc. (San Jose, CA, USA) for cooperation regarding the automation process and the Department of Urology (Medical University Innsbruck, Austria) for providing serum samples.

References

1. Petricoin EE, Paweletz CP, Liotta LA (2002) Clinical applications of proteomics: proteomic pattern diagnostics. J Mammary Gland Biol Neoplasia 7(4):433–440
2. Li J, Zhang Z, Rosenzweig J, Wang YY, Chan DW (2002) Proteomics and bioinformatics approaches for identification of serum biomarkers to detect breast cancer. Clin Chem 48(8):1296–1304
3. Gundry RL, Fu Q, Jelinek CA, Van Eyk JE, Cotter RJ (2007) Investigation of an albumin-enriched fraction of human serum and its albuminome. Proteomics Clin Appl 1(1):73–88
4. Tan CS, Ploner A, Quandt A, Lehtioe J, Pawitan Y (2006) Finding regions of significance in SELDI measurements for identifying protein biomarkers. Bioinformatics 22:1515–1523
5. Hutchens TW, Yip TT (1993) New desorption strategies for the mass-spectrometric analysis of macromolecules. Rapid Commun Mass Spectrom 7:576–580
6. Issaq HJ, Veenstra TD, Conrads TP, Felschow D (2002) The SELDI-TOF MS approach to proteomics: protein profiling and biomarker identification. Biochem Biophys Res Commun 292:587–592
7. Merchant M, Weinberger SR (2000) Recent advancements in surface-enhanced laser desorption/ionization time-of-flight mass spectrometry. Electrophoresis 21:1164–1177
8. Abramovitz M, Leyland JB (2006) A systems approach to clinical oncology: focus on breast cancer. Proteome Sci 4:1–15
9. Feuerstein I, Najam-ul-Haq M, Rainer M et al (2006) Material-enhanced laser desorption/ionization (MELDI) – a new protein profiling tool utilizing specific carrier materials for time of flight mass spectrometric analysis. J Am Soc Mass Spectrom 17(9):1203–1208
10. Najam-ul-Haq M, Rainer M, Trojer L, Feuerstein I, Vallant RM, Huck CW et al (2007) Alternative profiling platform based on MELDI and its applicability in clinical proteomics. Expert Rev Proteomics 4(4):447–452
11. Jemal A, Tiwari RC, Murray T et al (2004) Cancer statistics, 2004. CA Cancer J Clin 54(1):8–29
12. Oesterling JE (1991) Prostate specific antigen: a critical assessment of the most useful tumor marker for adenocarcinoma of the prostate. J Urol 145(5):907–923
13. Pannek J, Partin AW (1998) The role of PSA and percent free PSA for staging and prognosis prediction in clinically localized prostate cancer. Semin Urol Oncol 16(3):100–105
14. Feuerstein I, Rainer M, Bernardo K et al (2005) Derivatized cellulose combined with MALDI-TOF MS: a new tool for serum protein profiling. J Proteome Res 4(6):2320–2326
15. Trojer L, Lubbad SH, Bisjak CP, Wieder W, Bonn GK (2007) Comparison between monolithic conventional size, microbore and capillary poly(*p*-methylstyrene-*co*-1,2-bis(*p*-vinylphenyl) ethane) high-performance liquid chromatography columns: Synthesis, application, long-term stability and reproducibility. J Chromatogr A 1146(2):216–224

Chapter 6

Developing an iMALDI Method

Brinda Shah, Jennifer D. Reid, Michael A. Kuzyk, Carol E. Parker, and Christoph H. Borchers

Abstract

The iMALDI (immuno-MALDI) technique involves the affinity capture of target peptides from an enzymatic digest of a sample, followed by the direct analysis of the affinity beads while on a MALDI target. For determination of peptide concentration (and, by inference, protein concentration), stable-isotope-labeled standard peptides (SIS peptides) can be added to the digest and will be captured along with the native peptides. This technique can provide the highest possible specificity by determining two molecular characteristics of the epitope-containing peptides: (1) the molecular weight, typically measured to within 100 ppm or better by MALDI-MS, and (2) the amino acid sequence, by performing MALDI-MS/MS. This technique has been shown to be capable of detecting low-attomole levels of target peptides in environmental samples and in digests of human plasma. This chapter provides a detailed description of how to perform iMALDI analyses, starting with the selection of the target peptides. Examples are shown of the application of iMALDI to the detection of an organism that is a possible bioterrorism threat, and to the detection of two isoforms of human EGFR.

Key words iMALDI, Immuno-MALDI, MALDI, Immunoaffinity, Quantitation, EGFR, Francisella, EGFRvIII, Stable isotope labeling, SIS peptides

1 Introduction

iMALDI (from *i*mmunoaffinity and MALDI, matrix-assisted laser desorption/ionization) is a technique that involves affinity capture of peptides specific to a target protein. To make this technique quantitative, a known amount of a stable-isotope-labeled version of the peptide is added to the sample digest. An antibody, bound to the affinity beads, is used to capture both forms of the target peptide, whose amounts (in femtograms or femtomoles) can then be determined from the relative peak heights or peak areas of the labeled and the endogenous forms. One of the major advantages of iMALDI is to minimize losses due to sample handling.

In practice, many steps of this procedure are identical to those involved in epitope mapping, which was previously described in a previous volume of this series [1].

Helena Bäckvall and Janne Lehtiö (eds.), *The Low Molecular Weight Proteome: Methods and Protocols*, Methods in Molecular Biology, vol. 1023, DOI 10.1007/978-1-4614-7209-4_6, © Springer Science+Business Media New York 2013

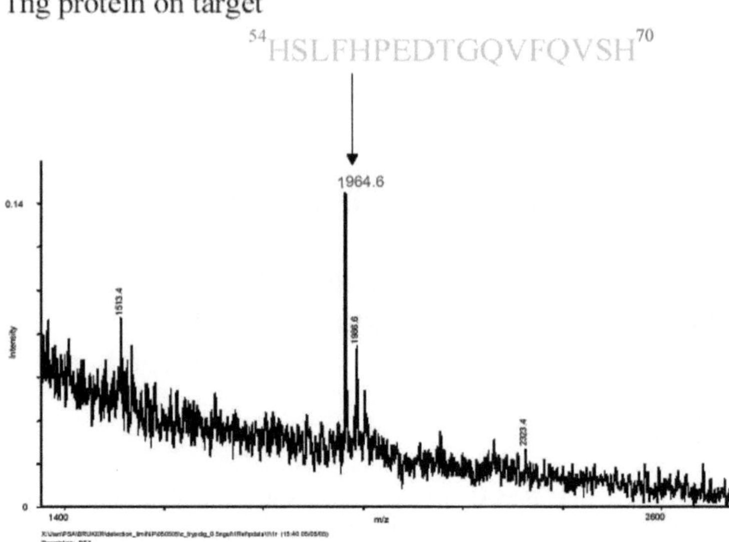

Fig. 1 MALDI-MS spectrum of epitope peptide (m/z = 1,964.6) from 1 ng PSA after digestion, affinity-bound to anti-PSA antibody beads

In fact, iMALDI is closely related to the epitope mapping technique—in the final steps of epitope mapping, what you have is affinity-bound peptides attached to antibody beads, which are placed onto the MALDI target for "direct analysis." In iMALDI, however, the proteins have already been digested into peptides *before* they are affinity captured. Co-capture of an isotopically labeled internal standard peptide makes this technique capable of absolute quantitation. Please note that in this chapter we are using the term "absolute" as it is commonly used in proteomics. Although the concentrations are still determined relative to an internal standard, in "absolute quantitation" the results are expressed in terms of an amount of material (i.e., a number of femtograms or femtomoles) in contrast to a "fold change".

To develop an iMALDI assay, one should first see if an antibody against a linear peptide from the target protein is already commercially available (*see* **Note 1**). If an antibody is commercially available, and the epitope is NOT known, then one could start by performing the epitope mapping to determine the target/epitope peptide. In one of our early iMALDI projects, for PSA, we started by determining the epitope peptide for PSA, which turned out to be an unexpected chymotryptic peptide (Fig. 1). Epitope mapping, however, is beyond the scope of this book chapter, but has been described previously [1].

The more common alternative, and the one we will describe in this book chapter, is to select a peptide based on its suitability for iMALDI analysis, and to have an antibody raised against this peptide. This is probably actually preferable, because it allows you to

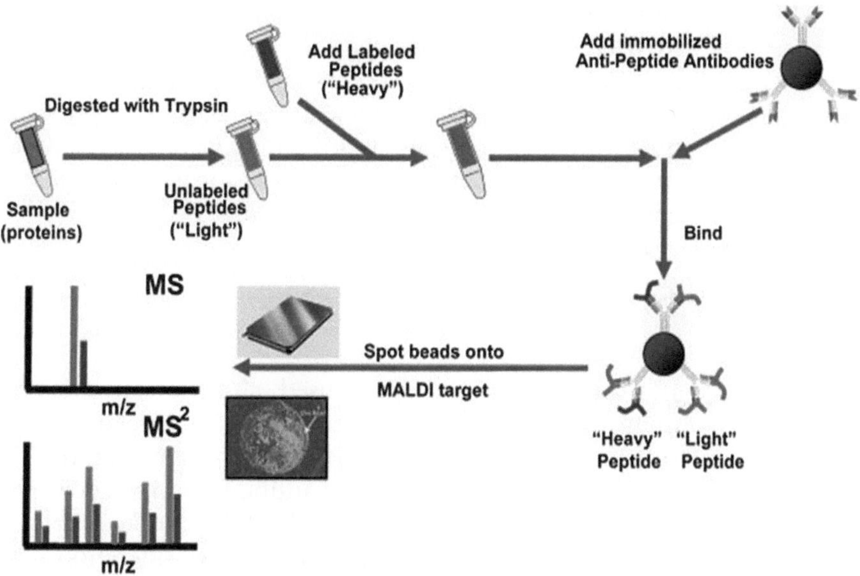

Fig. 2 Analytical scheme of an iMALDI assay. Reprinted from ref. 3, with permission

select a *high-sensitivity* target peptide. It can, however, be expensive because these peptides must be used to raise the antibodies. In practice, several high-sensitivity peptides that are specific to the target protein need to be selected and synthesized (*see* **Note 2**). Not all of these peptides will turn out to be antigenic—their binding affinities must be determined, and this can usually be done by the laboratory making the antibodies.

The methods described below illustrate the approaches used for several types of iMALDI projects. The first project described involved detection of a target peptide from a protein specific to a target organism, in this case, the 23 kDa protein, IGlC, from *Francisella tularensis* (*see* **Note 3**) [2]. The second type of project involved detection of a target peptide from human EGFR [3], and the third project involved detection of a target peptide from EGFR specific to a particular *variant* of human EGFR [4].

A "generic" iMALDI project is illustrated in Fig. 2. Although other enzymes could be used for digestion of the protein and the sample, for the sake of simplicity, we will assume that the digestion was done with trypsin. The epitope-containing tryptic peptides are synthesized using an isotopically labeled amino acid, and are added to a digest of the sample (plasma, tissue, etc.). These endogenous and labeled peptides are subsequently incubated with the antibody beads to immuno-adsorb both the isotopically labeled and the unlabeled peptides. After immuno-adsorption of the differentially labeled peptides, a small aliquot of the antibody beads can be arranged in a microarray/spot format on the MALDI-target plate. MALDI matrix solution added to the affinity-bound peptides elutes the peptides from the immobilized antibodies, permitting MALDI analysis of the peptides. The relative abundances of the molecular ion signals corresponding to "heavy" (isotopically labeled) and "light" (unlabeled, endogenous) peptides are used to quantify the amount of this protein in the original sample. Absolute specificity can be achieved by mass spectrometric sequencing of the epitope-containing peptide, using MALDI–MS/MS.

2 Materials

2.1 Obtaining an Antibody

2.1.1 Selection of Antibody (If Commercially Available)

If a commercial antibody is used, and the epitope is not known, then the epitope must be determined. For details of this process, which is beyond the scope of this book chapter, *see* ref. 1. If you are using a commercially available antibody, and the epitope is known, then this epitope peptide should still be checked for MALDI sensitivity (*see* Subheading 3.1). Another consideration is whether or not a tryptic peptide contains this epitope, as this will affect the choice of enzyme for digestion of the target tissue.

If you will *not* be using a commercially available antibody, then the first step is to select several target peptides that are sensitive in MALDI and are unique to the target protein or organism.

2.1.2 Selection of the Target Peptide if the Antibody Is Not Commercially Available

A MALDI-MS spectrum of a digested target protein, or target organisms, should be obtained (*see* Subheading 3.1). Examples of a "screening" MALDI-MS spectrum of a tryptic digest of the target proteins are shown in Fig. 3 for a digest of *F. tularensis* IglC, and in Fig. 4 for a digest of human EGFR. Several high-sensitivity peptides should be submitted for database searching to be certain that it is *specific* to the target protein (Fig. 5). In addition to sensitivity and specificity, factors such as amino acid composition should be considered. The ideal target peptide should probably NOT contain methionine, for example, because oxidation will lead to splitting the signal between several forms, making detection and quantitation more difficult. However, other considerations may take precedence: one of the peptides ideally selected for the *Francisella* project did contain methionine, and the assay was still successful. Likewise, an ideal endogenous peptide ideally should not contain cysteine because of the possibility of forming disulfide bonds with other peptides. Also, peptides containing missed cleavage sites are also not ideal choices for iMALDI peptides because their formation may not be reproducible (*see* **Note 4**).

2.1.3 Synthesis of the Peptide for Use in Raising the Antibody

The synthetic peptides usually correspond to the target sequence, with an additional N-terminal cysteine residue added so that it can easily be conjugated with carrier proteins [5]. This can be done by a peptide synthesis facility, but is also often offered as a service by

***Francisella tularensis* IglC sequence:**
1 MSEMITRQQVTSGETIHVRTDPTACIGSHPNCRLFIDSLTIAGEKLDKNIVAIEGGEDV 61 TKADSATAAASVIRLSITPGSINPTISITLGVLIKSNVRTKIEEKVSSILQASATDMKIK 121 KLGNSNKKQE YKTDEAWGIM IDLSNLELYP ISAKAFSISI EPTELMGVSK DGMSYHIISI 181 DGLTTSQGSL PVCCAASTDK GVAKIGYIAA A

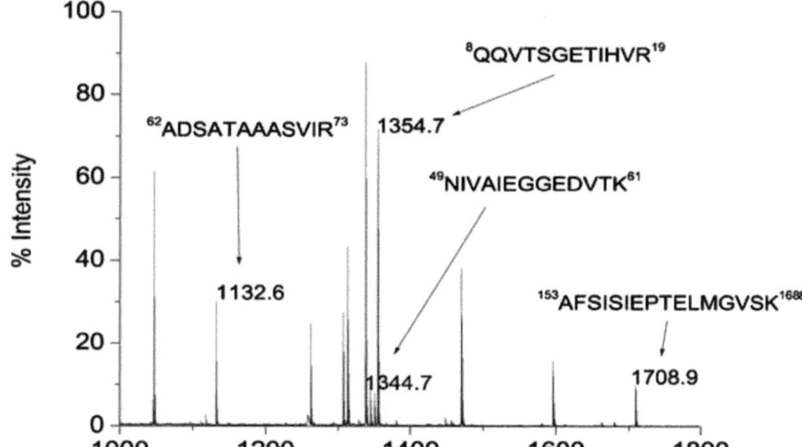

Fig. 3 Selection of *F. tularensis* IglC peptides for raising antibodies to be used for the *F. tularensis* iMALDI assay. MALDI-MS of proteolytic *F. tularensis* IglC peptides obtained by in-solution digestion of IglC with trypsin. Four tryptic peptides of IglC were selected for antibody production based on their high sensitivities in MALDI-MS. Reprinted from ref. 2, with permission

the group or company raising the antibody. It is usually possible for these laboratories to synthesize the peptide, raise the antibody, determine the binding affinity, and purify the antibody. These will be briefly discussed in the next few sections, although it is not expected that the mass spectrometry laboratory will be performing these steps.

2.1.4 Raising the Antibody

Raising an antibody is usually outside the scope of most mass spectrometry laboratories, and will not be discussed in detail in this book chapter. Fortunately, there are many commercial laboratories that will perform this service. We have used several laboratories, including SigmaGenosys (Woodlands, TX) for the *F. tularensis* work, Cocalico Biologicals (Reamstown, PA) and Immunoprecise Inc. (Victoria, BC) for the first EGFR project, and EZBiolab Inc. (Carmel, IN) for the EGFRvIII project.

Because of the costs involved, we have found it to be a useful strategy to raise a polyclonal antibody first. If the target peptide proves to be well suited for iMALDI, a monoclonal antibody can

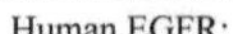

Human EGFR:

1 MRPSGTAGAA LLALLAALCP ASRALEEKKV CQGTSNKLTQ LGTFEDHFLS LQRMFNNCEV
61 VLGNLEITYV QRNYDLSFLK TIQEVAGYVL IALNTVERIP LENLQIIRGN MYYENSYALA
121 VLSNYDANKT GLKELPMRNL QGQKCDPSCP NGSCWGAGEE NCQKLTKIIC AQQCSGRCRG
181 KSPSDCCHNQ CAAGCTGPRE SDCLVCRKFR DEATCKDTCP PLMLYNPTTY QMDVNPEGKY
241 SFGATCVKKC PRNYVVTDHG SCVRACGADS YEMEEDGVRK CKKCEGPCRK VCNGIGIGEF
301 KDSLSINATN IKHFKNCTSI SGDLHILPVA FRGDSFTHTP PLDPQELDIL KTVKEITGFL
361 LIQAWPENRT DLHAFENLEI IRGRTKQHGQ FSLAVVSLNI TSLGLRSLKE ISDGDVIISG
421 NKNLCYANTI NWKKLFGTSG QKTKIISNRG ENSCKATGQV CHALCSPEGC WGPEPRDCVS
481 CRNVSRGREC VDKCNLLEGE PREFVENSEC IQCHPECLPQ AMNITCTGRG PDNCIQCAHY
541 IDGPHCVKTC PAGVMGENNT LVWKYADAGH VCHLCHPNCT YGCTGPGLEG CPTNGPKIPS
601 IATGMVGALL LLLVVALGIG LFMRRRHIVR KRTLRRLLQE RELVEPLTPS GEAPNQALLR
661 ILKETEFKKI KVLGSGAFGT VYKGLWIPEG EKVKIPVAIK ELREATSPKA NKEILDEAYV
721 MASVDNPHVC RLLGICLTST VQLITQLMPF GCLLDYVREH KDNIGSQYLL NWCVQIAKGM
781 NYLEDRRLVH RDLAARNVLV KTPQHVKITD FGLAKLLGAE EKEYHAEGGK VPIKWMALES
841 ILHRIYTHQS DVWSYGVTVW ELMTFGSKPY DGIPASEISS ILEKGERLPQ PPICTIDVYM
901 IMVKCWMIDA DSRPKFRELI IEFSKMARDP QRYLVIQGDE RMHLPSPTDS NFYRALMDEE
961 DMDDVVDADE YLIPQQGFFS SPSTSRTPLL SSLSATSNNS TVACIDRNGL QSCPIKEDSF
1021 LQRYSSDPTG ALTEDSIDDT FLPVPGEWLV WKQSCSSTSS THSAAASLQC PSQVLPPASP
1081 EGETVADLQT Q

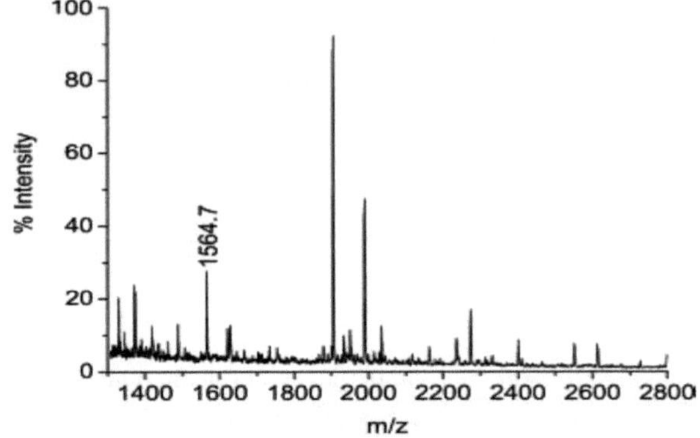

Fig. 4 Selection of EGFR peptides to be used for raising antibodies for the EGFR iMALDI assay. MALDI-MS of proteolytic EGFR peptides, obtained by in-solution digestion of EGFR with trypsin. The tryptic peptide ^{942}MHLPSPTDSNFYR^{954} from EGFR (*m/z* 1,564.7) (sequence shown in *inset*) was selected for antibody production based on its high sensitivity in the MALDI-MS mode. Reprinted from ref. 3, with permission

then be raised against the same synthetic target peptide. Antibodies are usually raised in rabbits (at least three rabbits per peptide), and the process takes 77 days. Several bleeds are collected, and the antibodies are purified from the blood using the same peptide that was used to raise the antibody—except that these peptides have been immobilized. For binding the antibody to the beads, the final buffer that the antibody is provided in should NOT contain amine groups, as this will interfere with the coupling of the antibody to the bead. We recommend using PBS, if possible.

2.1.5 Determining the Binding Affinity of the Antibody

An ideal iMALDI antibody should exhibit high-affinity binding, as measured by the ELISA response factor (Table 1). This is usually also determined by the contract laboratory. An ELISA response factor of >100,000 is considered to be a high-titer animal. If an antibody does not exhibit high binding efficiency, it is not worth

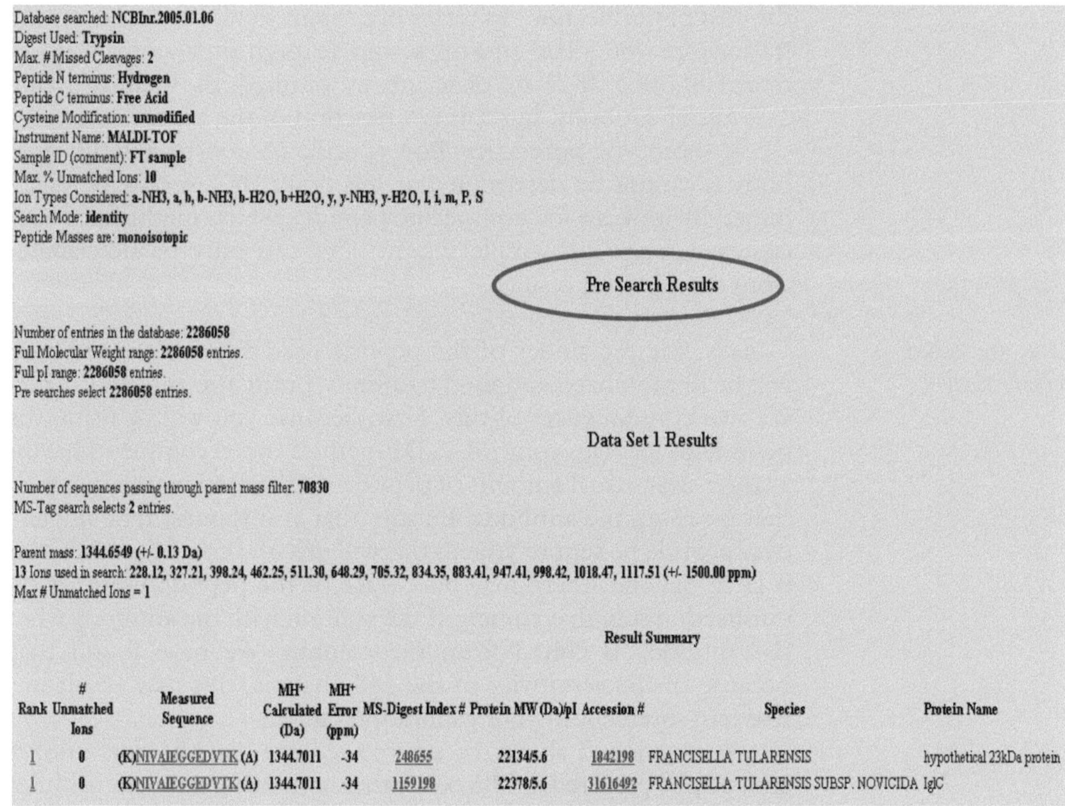

Database searched: **NCBInr.2005.01.06**
Digest Used: **Trypsin**
Max. # Missed Cleavages: **2**
Peptide N terminus: **Hydrogen**
Peptide C terminus: **Free Acid**
Cysteine Modification: **unmodified**
Instrument Name: **MALDI-TOF**
Sample ID (comment): **FT sample**
Max. % Unmatched Ions: **10**
Ion Types Considered: **a-NH3, a, b, b-NH3, b-H2O, b+H2O, y, y-NH3, y-H2O, I, i, m, P, S**
Search Mode: **identity**
Peptide Masses are: **monoisotopic**

Pre Search Results

Number of entries in the database: **2286058**
Full Molecular Weight range: **2286058** entries.
Full pI range: **2286058** entries.
Pre searches select **2286058** entries.

Data Set 1 Results

Number of sequences passing through parent mass filter: **70830**
MS-Tag search selects **2** entries.

Parent mass: **1344.6549 (+/- 0.13 Da)**
13 Ions used in search: **228.12, 327.21, 398.24, 462.25, 511.30, 648.29, 705.32, 834.35, 883.41, 947.41, 998.42, 1018.47, 1117.51 (+/- 1500.00 ppm)**
Max # Unmatched Ions = 1

Result Summary

Rank	# Unmatched Ions	Measured Sequence	MH+ Calculated (Da)	MH+ Error (ppm)	MS-Digest Index #	Protein MW (Da)/pI	Accession #	Species	Protein Name
1	0	(K)NIVAIEGGEDVTK(A)	1344.7011	-34	248655	22134/5.6	1842198	FRANCISELLA TULARENSIS	hypothetical 23kDa protein
1	0	(K)NIVAIEGGEDVTK(A)	1344.7011	-34	1159198	22378/5.6	31616492	FRANCISELLA TULARENSIS SUBSP. NOVICIDA	IglC

Fig. 5 BLAST search of the target peptide from *Francisella tularensis*. There is only one hit in the NCBInr "all-species" database, showing that his peptide is unique to *F. tularensis*

Table 1
Affinity determination of the four *Francisella tularensis* anti-IglC-peptide antibodies. Reprinted from ref. 2, with permission

F. tularensis IglC peptides used for antibody production	Anti-*F. tularensis* IglC peptide antibodies	ELISA response (titer)
^{153}AFSISIEPTELMGVS^{167}	[(*F. tularensis* IglC) aa153–167]	1/50,000
^{8}QQVTSGETIHVR^{19}	[(*F. tularensis* IglC) aa8–19]	1/3,000
^{48}C*NIVAIEGGEDVTK^{61}	[(*F. tularensis* IglC) aa48–61]	1/500,000
^{62}ADSATAAASVIR^{73}	[(*F. tularensis* IglC) aa62–73]	1/10,000

The cysteine residue marked with asterisk has been added to the native sequence because of ease of conjugation with the carrier protein the cost of purification. In terms of protein yield, antibody concentrations can be ~100 μg/ml serum in high-titer animals (more than 100,000). If 25 ml of serum are purified, the yield is usually 1–2 mg of antibody, but this is a function of the titer.

It should be noted that how specific or nonspecific this antibody is cannot be determined at this point. It can only be determined by looking for nonspecific peptides which might have been captured from the sample digest. This can only be determined from the MALDI results.

2.1.6 Purification of the Antibody

Usually, the remainder of the peptide used to raise the antibody will be immobilized and used to affinity-purify the antibody. There are two consequences of this. First, because you will be using this peptide to validate your iMALDI method, we recommend having a larger-than-usual amount of peptide synthesized by the company that is raising the antibody. Be sure that you request that any leftover peptide be sent to you, as this will save the cost of synthesizing it later. Second, there may be a trace of the peptide used for the purification step that comes off the column with the antibody when the antibody is eluted from the column. We have found that, because of the sensitivity of the MS analysis, this may result in a nonzero intercept in your calibration curve.

As mentioned above, be sure to request that the final purified antibody be prepared in PBS or another non-amine-containing buffer, to avoid problems in coupling the antibody to the beads. Antibodies in a BSA solution, for example, will result in the preparation of immobilized BSA beads and not immobilized antibody beads.

2.2 Reagents and Solvents

Please note that other reagents and suppliers may be substituted for those listed here, and may provide equally good results, but we have had no experience with their use. In each case where a particular manufacturer is listed, please recognize that the phrase "or the equivalent" is implied.

2.2.1 Preparation and Digestion of Standard Protein (Target Protein)

1. 50 mM ammonium bicarbonate (50 mM ABC solution).
2. Trypsin, porcine, sequencing-grade modified (Promega).

2.2.2 Preparation and Digestion of Sample

1. 25 mM ammonium bicarbonate (25 mM ABC solution).
2. Dithiothreitol (DTT).
3. Iodoacetamide.
4. Sodium deoxycholate (NaDOC).
5. Trypsin, porcine, sequencing-grade, modified (Promega).
6. Oasis® HLB Extraction Cartridge (Waters Corporation, Milford, MA), or the equivalent.

2.2.3 Preparation of CNBr-Activated Sepharose Beads, Direct Coupling of the Primary Antibody to the Beads, and Immunoprecipitation of Peptides

1. CNBr-activated Sepharose 4B Beads (GE Healthcare Life Sciences, Piscataway, NJ, 17-0430-01).
2. HCl solution: 1 mM hydrochloric acid (HCl) in HPLC-grade water (*see* **Note 5**).
3. Compact Reaction Columns (CRC) (USB, Cleveland, OH); columns: 13928; 35-μ compact column filters, 13912 (*see* **Note 6**).
4. Coupling buffer: 0.1 M sodium bicarbonate, pH 8.3, 0.5 M NaCl.
5. Blocking buffer: 0.1 M Tris–HCl (Trizma Hydrochloride), pH 8.0.
6. Acetate buffer: 0.1 M sodium acetate, 0.5 M NaCl, pH 4.0.

2.2.4 Preparation of Magnetic Protein-G Beads, for Indirect Coupling of the Primary Antibody to the Beads, and Immunoprecipitation of Peptides

1. CHAPS solution: 1× PBS (*see* **Note 7**) with 0.3 % CHAPS.
2. PBS solution: 1× PBS, pH 7.2.
3. Ammonium bicarbonate (ABC).
4. Dynabeads® Protein-G (Invitrogen Life Technologies, 100-03D).
5. DynaMag®-2 Magnet (Invitrogen Life Technologies, 12321D), or the equivalent.

2.3 MALDI Matrix, MALDI Matrix Solvent, and Mass Calibration Standards

1. MALDI matrix solvent: 1.8 mg/ml ammonium citrate in 70 % ACN, 0.1 % trifluoroacetic acid (TFA).
2. MALDI matrix solution: 3 mg/ml α-cyano-4-hydroxycinnamic acid in MALDI matrix solvent.
3. MALDI mw standards:
    - Bovine serum albumin (BSA), lyophilized powder, fatty acid free, globulin free, ≥99 %, purified by gel electrophoresis.
    - Carbonic anhydrase I, powder, from human erythrocytes, powder.
    - Insulin B oxidized, powder, from bovine pancreas.
    - Angiotensin 1 (human), acetate salt ≥90 %, HPLC purified.

2.4 MALDI Instrumentation

The projects described here were performed over the course of several years and in different laboratories. The first MALDI instruments used for beads-on-the-target research projects were the Perceptive Voyager RP, a Perseptive DE-STR, and a Bruker Reflex I. Since then, a Bruker Ultraflex I, an Applied Biosystems Q-Star Pulsar, and an Applied Biosystems 4800 MALDI-TOF/TOF have been used. Some instrument manufacturers, however, do not recommend the use of affinity beads directly on the target, because of the possibility of the laser dislodging these beads which can then be caught on screens inside the vacuum system. For example, the 4700 series of the Applied Biosystems MALDI-TOF/TOF is not recommended for beads-on-the-target experiments. We highly recommend checking with the manufacturer of your MALDI instrument before performing these experiments (*see* **Note 8**).

3 Methods

3.1 Digestion of Standard Protein

1. Digest the target protein with trypsin (or use the extract from an in-gel-digested protein instead (*see* **Note 9**)).
2. Dissolve one 20 μg vial of trypsin in 200 μl 50 mM ABC buffer.
3. Dissolve 1 μg target protein in 20 μl of 50 mM ABC buffer.
4. Add 1 μl trypsin solution.
5. Digest overnight at 37 °C.
6. Prepare a fresh solution of MALDI matrix (α-cyano-4-hydroxy-cinnamic acid, CHCA) in MALDI matrix solvent (*see* **Note 10**). CHCA matrix is ideal for most peptides (*see* **Note 11**).
7. Spot 0.3 μl of digest.
8. Spot 0.3 μl of MALDI matrix solution onto sample, and allow to air-dry. Analyze by MALDI-MS.
9. Examine peptides to determine sensitivity and specificity, as well as suitability for iMALDI analysis.
10. Do a BLAST search on the peptide sequence (http://blast.ncbi.nlm.nih.gov/Blast.cgi) to determine that the peptide is specific to the target protein or organism.

Our most recent iMALDI project involved the detection and quantitation of variant EGFR versus normal EGFRvIII [4]. These two protein isoforms involve deletion of exons 2–7 and insertion of a glycine (Fig. 6). In this case, the selection of the target peptide was dictated by the project: the two variants produce a peptide that differs by one amino acid (K)**G**NYVVTDHGSCVR versus (R)NYVVTDHGSCVR.

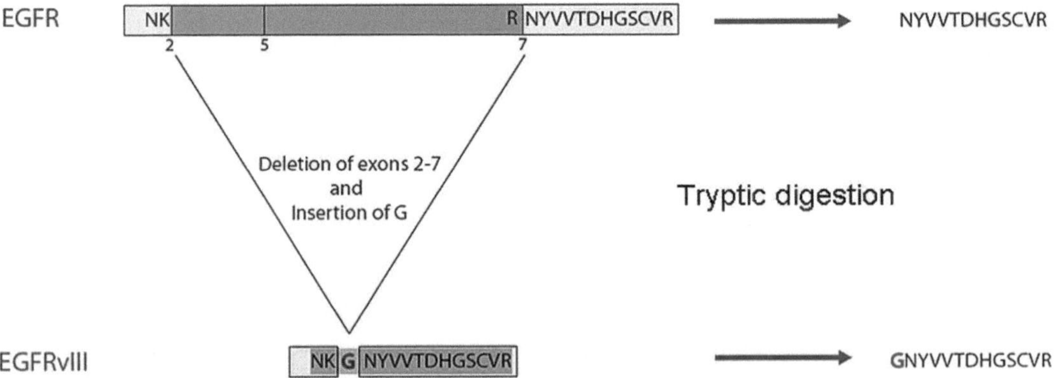

Fig. 6 The difference between normal human EGFR and EGFRvIII. Reprinted from ref. 4, with permission

3.2 Coupling the Antibody to the Beads

For direct coupling of the primary antibody to the beads, proceed to Subheading 3.2.1. For indirect coupling of the primary antibody, proceed to Subheading 3.2.3.

3.2.1 Preparation of CNBr-Activated Sepharose-Immobilized Antibody Beads for Direct Coupling

CNBr-activated Sepharose beads are prepared according to the procedures provided with the packing material, but on a much smaller scale. Briefly, the Sepharose beads react with primary amine groups on the antibody, covalently linking the antibody to the beads. Unreacted binding sites on the beads are blocked by incubation with an amine-containing blocking buffer (here, Tris buffer), and the process is completed by performing a series of washes at alternating acidic and basic pH using the acetate and coupling buffers.

1. Put approximately 0.2 g of dry CNBr Sepharose beads into a Falcon tube.
2. Add 10 ml of HCl solution.
3. Swirl the mixture and let equilibrate for 15 min.
4. Add approximately 100 μl of the wet beads from the bottom of the tube to each of two compact reaction columns (CRCs) (*see* **Notes 12** and **13**), and drain the columns.
5. Wash the CNBr-activated Sepharose mini-column with 6 × 0.8 ml with HCl solution.
6. Store wet at 4 °C until used, in accordance with the manufacturer's instructions.

3.2.2 Direct Coupling: Binding the Antibody Directly to CNBr Sepharose Affinity Beads

Binding the antibody to the beads should be done at a temperature and pH where the antibody is stable. These conditions used for this coupling step depend on the stabilities of both the protein and the antibody. Usually this is done either at 4 °C overnight or at room temperature for 1–2 h (*see* **Note 14**).

1. Thaw the antibody solution (which is usually shipped and stored on dry ice) immediately before use. Avoid unnecessary freeze–thaw cycles. Check the buffer that the antibody is shipped in. If it is an amine buffer, it will interfere with the coupling reaction, and the buffer solution will have to be exchanged before using.
2. Incubate the columns with the antibody solution (~20–50 μg antibody) in the coupling buffer for 2 h with slow rotation to allow coupling of the protein to the antibody. To avoid bead breakage, do not use a magnetic stirrer.
3. Wash with 5 column volumes of coupling solution (*see* **Note 15**).
4. Block any unreacted sites by rinsing with 0.4 ml of blocking solution.
5. Rotate at room temperature for 2 h.

6. Remove any remaining unbound antibody with a series of 5 bead-volume washes, alternating between the acetate and the coupling buffers. Add enough buffer solution to keep the beads moist at all times (*see* **Note 15**).
7. Proceed to Subheading 3.4.

3.2.3 Indirect Coupling: Preparation of Protein G Beads for Indirect Coupling Through a Secondary Antibody

Protein-G immobilized magnetic beads were used for our recent EGFRvIII project [4]. These beads offer several advantages over CNBr Sepharose beads. The preparation time is reduced because the anti-peptide antibody does not need to be covalently linked to the beads, and the use of magnets to manipulate the movement of the beads is much easier than the use of columns. These beads are also more uniform in size and shape than the Sepharose beads.

1. Unfortunately, however, the commercially available magnetic Protein-G beads we used are only available in a Tween20-containing storage buffer. The presence of this detergent prevents the beads from sticking to the walls of the sample tube, but causes a great deal of noise on the MALDI spectrum. As a result, the detection of low-abundance peptides is difficult if high quantities of Tween are present in the sample. In this case, the Tween20 buffer is exchanged with a CHAPS buffer, which shows a single peak on the spectrum as opposed to the several polymer peaks of Tween20 (*see* **Note 16**).
2. Prepare PBS-CHAPS solution.
3. Pipette 5 μl of bead slurry into 200 μl of PBS-CHAPS solution and incubate for 5–10 min at room temperature on slow rotation (*see* **Note 17**).
4. Using a magnet, attract the beads to one side of the sample tube and remove the supernatant. Replace with another 200 μl of the PBS-CHAPS solution.
5. Repeat at least five times to completely exchange the Tween20 detergent for CHAPS.
6. Resuspend the beads in 5 μl of pH 7.2 PBS solution (*see* **Note 18**).

3.2.4 Binding the Primary Antibody to the Protein-G Beads

1. Incubate 5 μl of the washed bead slurry with 1 μg of antibody in 100 μl of 1× PBS, pH 7.2, at room temperature on slow rotation for 2 h (*see* **Note 15**). This step can be performed in conjunction with the procedures outlined in Subheadings 3.5 and 3.6 (*see* **Note 19**).
2. Prepare separate sample tubes for different peptide solutions and controls.

3.3 Synthesis of the Labeled Version of the Target Peptide

A "heavy" form of the *F. tularensis* IglC peptide ^{49}NIVAIEGGED <u>V</u>TK^{61} was synthesized to contain an isotopically labeled valine at position 59 (underlined). The "heavy" peptide, 6 Da heavier than the endogenous isoform, was synthesized at the UNC Peptide Synthesis Facility, using ^{13}C-labeled fmoc valine purchased from Isotec/Sigma–Aldrich according to the fmoc approach described in details elsewhere [6].

The EGFR standard isotopically labeled peptide was ^{942}MHLPSPTDSNFYR^{954}, with isotopically labeled leucine at position 965 (underlined). This leucine was synthesized with ten deuteriums in place of ten hydrogens, resulting in a peptide 10 Da heavier than the endogenous form. The EGFRvIII standard peptide was GNYVVTDHGSCVR, where R was isotopically labeled with ^{13}C and ^{15}N.

3.4 Preparation and Digestion of the Sample

As is the case in *any* method of protein quantitation where a peptide is used to determine the concentration of the protein in the original sample, care must be taken to perform the digestion as completely and reproducibly as possible. Although quantitation using known amounts of stable-isotope-labeled standards is called "absolute" quantitation, it is actually still relative—relative to the internal standard. This type of quantitation can, however, be quite close to absolute, if the digestion is complete. Reproducibility is another factor—any variability in the digestion efficiency will cause inaccuracies and variability in the quantitation results.

Our previous methods have involved the use of an ~1:10 enzyme:substrate ratio of trypsin (sequencing-grade modified Promega porcine trypsin) to sample protein, and digestion of human plasma (UNC blood bank) was carried out in 25 mM ammonium bicarbonate (Sigma) at 37 °C overnight [3]. For the digestion of tissue, however, a more rigorous procedure may be required [7]. If the target protein has disulfide bonds, recommended methods include reduction and alkylation using DTT and iodoacetamide. A study recently performed in our laboratory indicates that a method using deoxycholate (DOC) for protein denaturation and membrane protein solubilization provides improved efficiency and reproducibility, which are absolutely critical for accurate quantitation [8].

1. Dissolve protein sample in 25 mM ABC.
2. Add NaDOC up to 1 % w/v.
3. Reduce with 1 μg of DTT per ~50 μg of sample protein at room temperature or 37 °C for 30 min.
4. Alkylate with 5 μg of iodoacetamide per approximately 50 μg of sample protein at room temperature or 37 °C for 1 h.
5. Digest with 1 μg of trypsin per approximately 50 μg of sample protein at room temperature or 37 °C overnight (*see* **Note 20**).

3.5 Adding the Synthetic-Labeled Peptide to the Sample

For absolute quantitation of the endogenous target peptide, a known concentration of the synthetic-labeled peptide should be spiked in (*see* **Note 21**). To create a standard curve, the labeled ("heavy") peptide is kept constant while the concentration of the target sample is varied and signal intensity ratio with MALDI-MS is measured. This curve is used to measure the amount of endogenous peptide in subsequent samples. In our laboratory experiments, we have found that adding the heavy peptide following the digestion of the sample yields more reliable results [8]. Additionally, to avoid quantitation errors, diluted sample should be used to generate the standard curve.

3.6 Incubation of the Sample with the Affinity Beads

1. Incubate one of the columns containing antibody beads with the sample digest for 2 h at room temperature, with slow rotation. Incubate the second column with PBS to serve as a control.
2. Wash the columns three times with 0.4 ml PBS, pH 7.2, to remove any unbound products, and remove an aliquot for MALDI analysis (*see* **Note 21**).

3.7 Placing the Beads on the Target

1. Using a pipette tip with the tip cut off, transfer an aliquot of the bead slurry containing approximately 10–15 beads to the MALDI target.
2. Prepare a 3 mg/ml α-cyano-4-hydroxycinnamic acid (CHCA) MALDI matrix in MALDI matrix solvent (*see* **Note 22**). CHCA matrix is ideal for most peptides (*see* **Note 23**).
3. Spot 0.3 μl of MALDI matrix onto sample, and allow to air-dry.

3.8 MALDI Analysis

External mass calibration should be performed, using two points that bracketed the mass range of interest. The MALDI instrument uses a stainless-steel target, on which the samples are deposited and dried. Instruments are equipped with a video camera, which displays a real-time image on a monitor, and the laser can be aimed at specific features within the area of the target. When the matrix solution dries, the beads shrink and stick to the target. However, it is better to not aim the laser directly at the affinity beads to avoid the possibility of "popping" them with the laser, even though it rarely happens.

Relative signal abundances in the MALDI spectra are not directly related to the relative abundances of different peptide species, due to potential differences in the sensitivities of different peptides. However, the response factors for different isotopically labeled forms of the *same* peptide should be identical, which allows quantitation to be done on the affinity-captured peptides. Standard MALDI-MS is used for the determination of potential target peptides. For additional confidence in the identification, MALDI-TOF/TOF instruments can be used to obtain MS/MS sequence information on the affinity-bound peptide.

Figures 7, 8, and 9 show the pairs of peaks resulting from the affinity capture and iMALDI analysis of the endogenous peptides

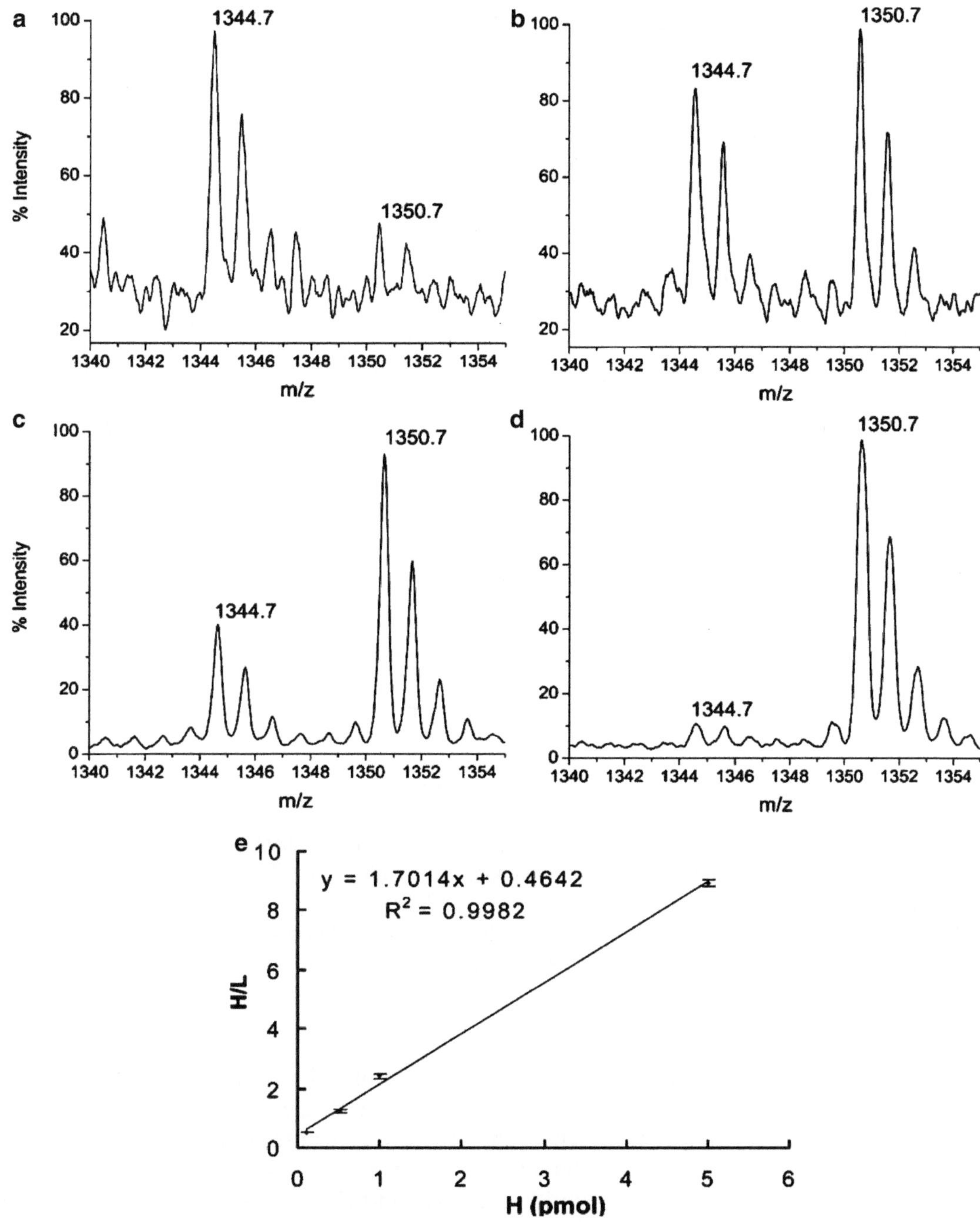

Fig. 7 Quantitation of *Francisella tularensis* bacteria using the *F. tularensis* iMALDI assay. Absolute quantitation of the IglC peptide NIVAIEGGEDVTK (aa49–61) (L, light peptide, *m*/*z*=1,344.7) in a bacterial sample. *F. tularensis* bacteria were digested and incubated with different amounts of heavy peptides (H, *m*/*z*=1,350.7) as internal standards: (**a**) 0.1 pmol, (**b**) 0.5 pmol, (**c**) 1 pmol, (**d**) 5 pmol, (**e**) plot of the observed ratios of monoisotopic abundances of H and L in the MALDI–MS spectra (**a**–**d**) versus the absolute amount of H added. Only a 10 % aliquot of the extract has been used for the analysis. Reprinted from ref. 2, with permission

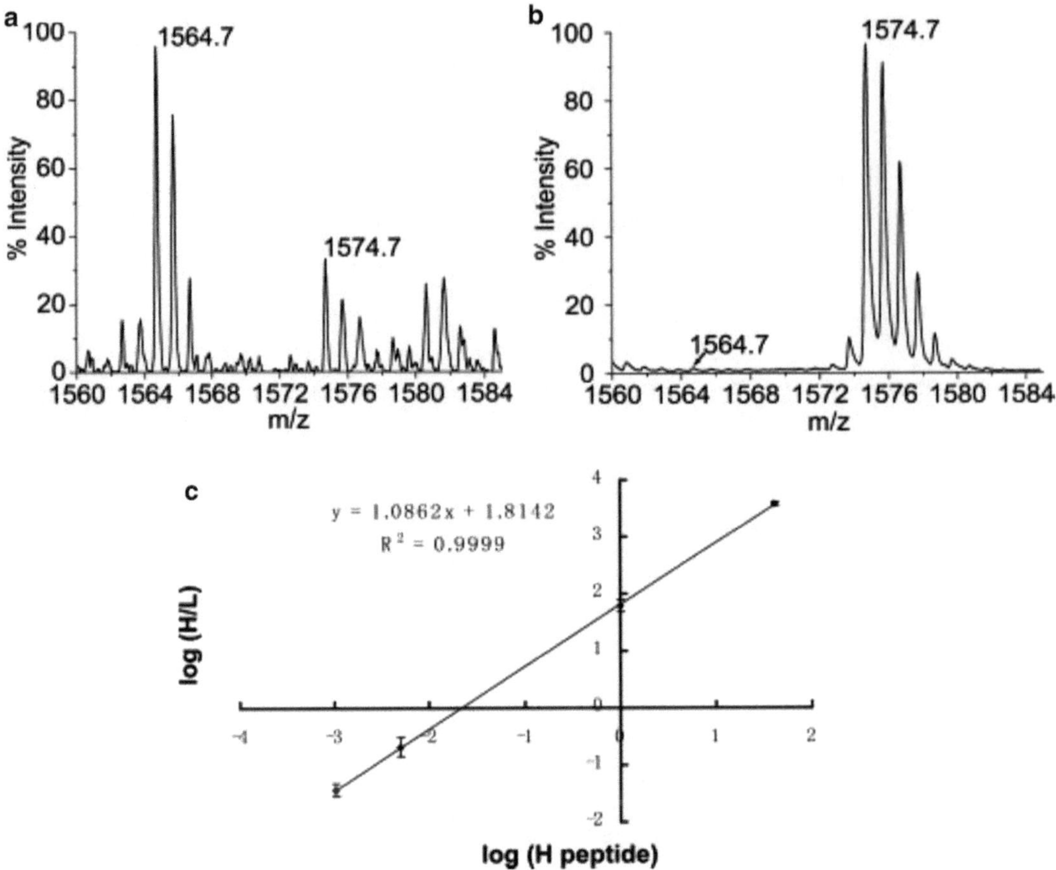

Fig. 8 Quantitation of EGFR in SUM102 cells using the EGFR iMALDI assay. Absolute quantitation of the EGFR peptide MHLPSPTDSNFYR (aa 963–975) (L light peptide, $m/z = 1,564.7$) in SUM102 cells. Cells were lysed, digested, and incubated with known amounts of heavy peptides (H, $m/z = 1,574.7$) as internal standards: (**a**) 0.1 pmol, (**b**) 1 pmol of H. (**c**) Logarithmic plot of the observed ratios of monoisotopic ion abundances of H and L in the MALDI-MS spectra versus the absolute amount of H added. Reprinted from ref. 3, with permission

and their isotope-labeled versions. The bottom panel (c) in each figure shows the linearity of the peak areas, generating a "standard curve" from which the absolute concentration in the original sample can be determined. The on-bead detection limits for the *Francisella* peptide (Fig. 7c) and the EGFR peptide in the earlier study (Fig. 8c) were on the order of 100 fmol. In our most recent study on EGFRvIII, the linear calibration curve of the peptide that distinguishes EGFRvIII from the normal variant was linear down to approximately 1 fmol (Fig. 9b).

Figures 10 and 11 show MS/MS spectra of the affinity-captured target peptides for *Francisella tularensis* IglC and for the peptide unique to the EGFRvIII peptide (*see* **Note 24**). The MS/

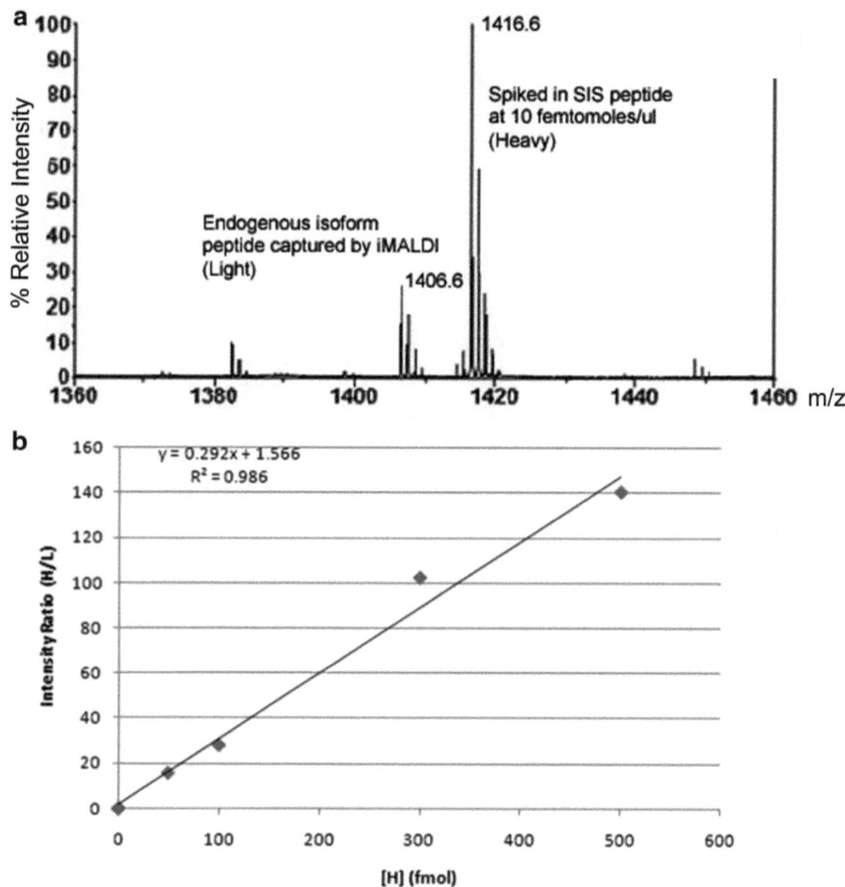

Fig. 9 Absolute quantitation of the endogenous isoform peptide from the tryptic digest of the U87MGΔEGFR cell line lysate. A stable-isotope standard (GNYVVTDDHGSCVR) was used "co-captured" and used as the internal standard. (**a**) iMALDI-MS spectrum showing co-captured "light" (endogenous) and "heavy" (SIS) peptides. (**b**) Absolute quantitation of the endogenous isoform. 1–500 fmol/μl of the "heavy" standard was spiked into a tryptic-digested sample to demonstrate the linear dynamic range of the assay. Reprinted from ref. 4, with permission

MS spectrum of the EGFRvIII peptide confirms the identity of this peptide as different from the wild type by the presence of the glycine-containing b_3 ion at $m/z = 335.08$.

iMALDI is clearly a sensitive method for the detection and quantitation of target proteins and peptides, combining affinity enrichment with specific detection. There are several advantages of this technique that may not be readily apparent. The first is that the simultaneous detection of the isotopically labeled standard peptide provides evidence that the entire assay (from the point of adding the standard peptide to the sample) is working. If there were no standard peptide, and no peak was detected during the analysis of the sample, there might be some question as to whether a problem

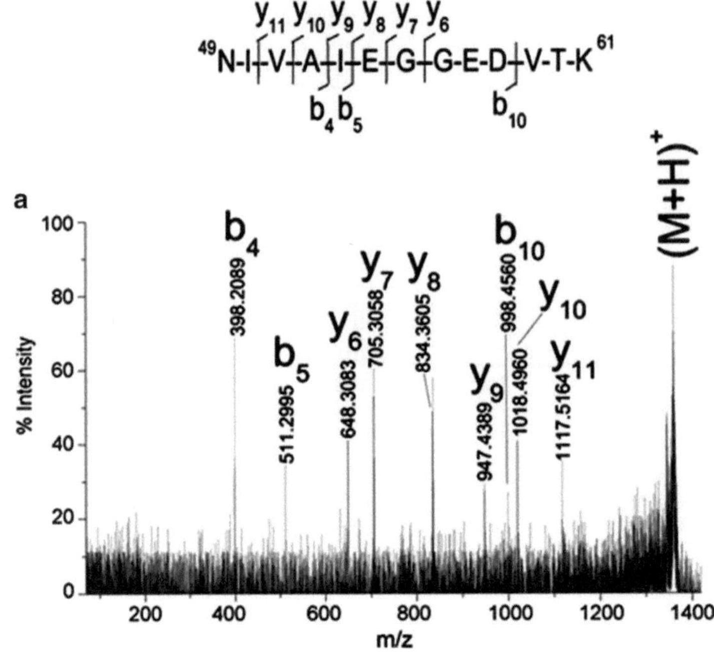

b

Fragment ions	m/z observed	m/z theoretical	Mass accuracy (ppm)
b_4	398.2089	398.24	-79
b_5	511.2995	511.32	-49
b_{10}	998.4560	998.48	-24
y_6	648.3083	648.32	-19
y_7	705.3058	705.34	-51
y_8	834.3605	834.38	-29
y_9	947.4389	947.47	-31
y_{10}	1018.4960	1018.51	-9.5
y_{11}	1117.5164	1117.57	-52

Fig. 10 *Francisella tularensis* iMALDI-MS/MS spectrum. (**a**) MALDI–MS/MS spectrum of the peptide at $m/z = 1344.7$ affinity-bound to anti-aa49–61 (*F. tularensis* IglC) antibody beads, obtained after proteolysis of *F. tularensis* bacteria in PBS solution. Reprinted from ref. 2, with permission

had occurred during the process or during the MALDI analysis. The detection of the SIS peptide eliminates this possibility.

Second, while these projects involved the detection of single peptides representing single proteins, we envision the production of protein "chips," targets printed with multiple types of beads which could be used to detect a panel of biomarkers for a single disease or multiple diseases. Beads containing different affinity-bound antibodies could be incubated with a biological fluid in a 96-well format, and a few beads from each well could be transferred to a MALDI target to perform these analyses in an automated manner. The liquid-handling devices capable of performing this

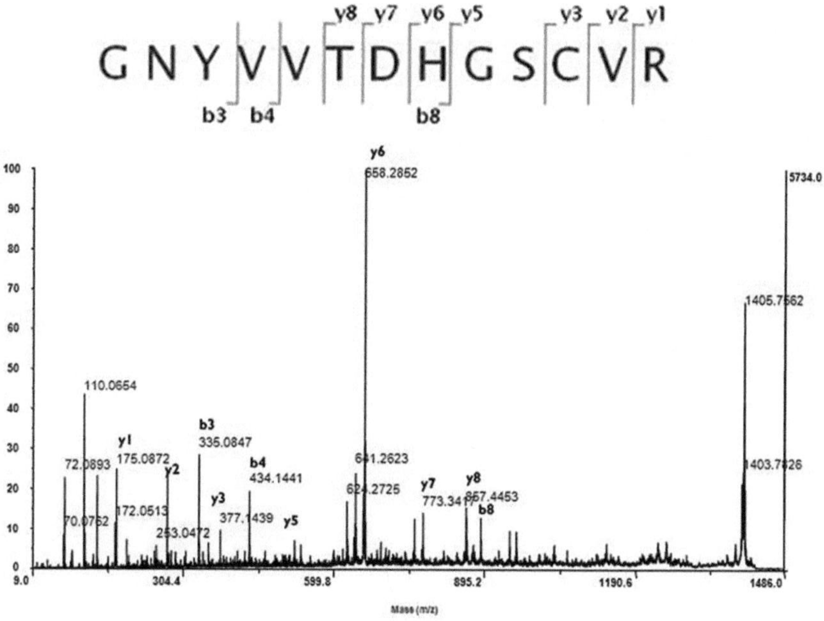

Fig. 11 iMALDI-MS/MS spectrum of the EGFRvIII peptide which contains an added glycine at the N-terminus compared to the wild-type peptide. Reprinted from ref. 4, with permission

transfer have already been demonstrated (Fig. 12). The automation of this technique could make the low-level detection of biomarkers a truly high-throughput technique.

4 Notes

1. The epitope should be linear—i.e., the epitope-containing peptide should be a continuous stretch of amino acids. Otherwise, denaturing and digesting the sample will denature the epitope peptide and it will not bind to the antibody. If an antibody works in Western blot analysis, it must have a continuous epitope because the protein is denatured during the assay.
2. The sensitivity should be determined in the MALDI mode because the sensitivities of peptides in MALDI may not be identical to the sensitivities in ESI, and vice versa.
3. This was the approach used for the iMALDI project on the detection of *Francisella* bacteria [2]. In the case of developing a detection method for an entire organism, as we did for *Francisella tularensis*, the entire organism *could* have been digested because *any* sensitive and specific peptide from *any Francisella* protein could have been used as the target peptide.
4. The size of the peptide is also a factor—ideally, the peptide will have a molecular weight between 1,000 and 2,500 Da, the

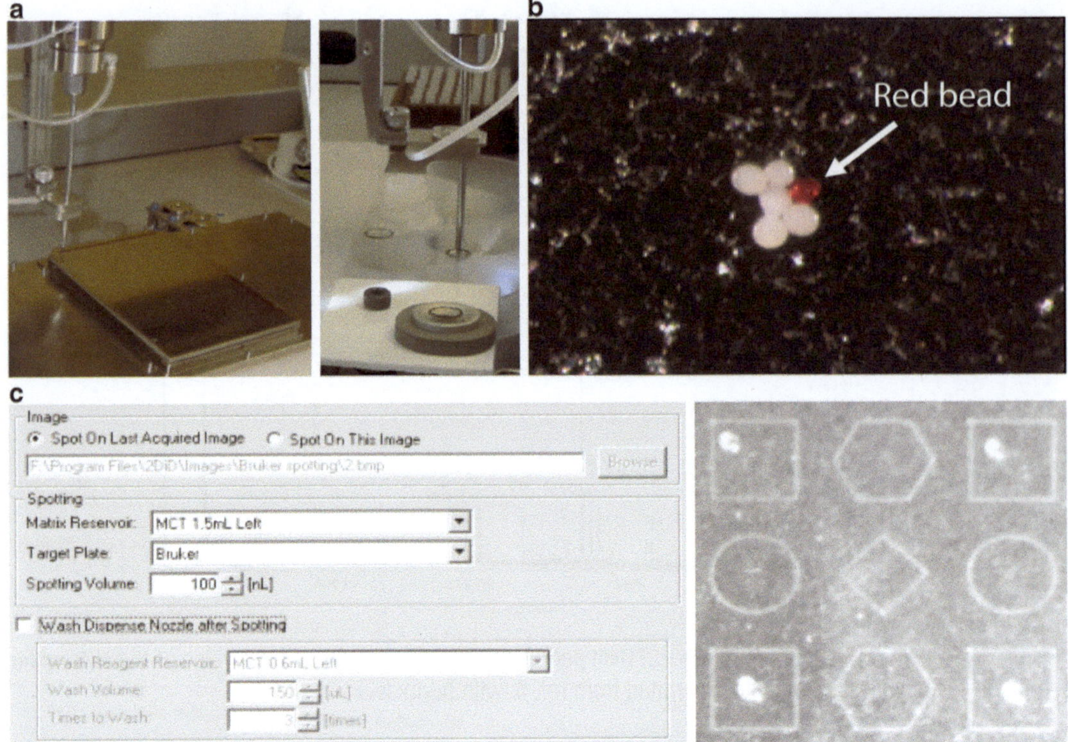

Fig. 12 Automation of an iMALDI experiment. (**a**) Automatic spotting of antibody beads directly onto MALDI target using 2DiD robotics system from BioMachines. Antibody beads in a 100 nl volume were spotted directly onto MALDI target from (**b**) ABI4700 and from (**c**) Bruker Daltonics (AnchorChip)

ideal range for MS/MS sequencing. For additional peptide selection criteria, *see* refs. 8, 9.

5. Unless otherwise stated, all of the solutions in this chapter are made with HPLC-grade water.
6. Be sure to order lower filters with these columns. They are not included with the sets, which include the bodies, plugs, and caps.
7. We usually make a 10× concentrated PBS solution, which can be sterilized and stored for several months at room temperature. A recipe for PBS can be found at http://www.thelabrat.com/protocols/4.shtml.
8. Since the analytes are actually eluted from the beads by the matrix solvent, it should, in theory, be possible to use the matrix solvent to desorb the analytes from the beads, and spot only the supernatant, albeit this procedure could lead to a decrease in the absolute sensitivity of the technique due to adsorption onto the walls of the container. Alternately, one could try physically removing the beads from the target after the matrix has been spotted and dried. However, we have not

actually tried these alternative procedures and do not recommend them because this would eliminate the major advantage of iMALDI, which is to minimize losses due to sample handling. We only include these suggestions in case you have an instrument that is VERY sensitive to particles.

9. If your target protein is found on a gel, and was identified using in-gel digestion [10], the MALDI-MS spectrum of this digest is another way of determining the MALDI sensitivities of the proteolytic peptides from this protein.
10. Better sensitivity is sometimes obtained if the α-cyano-4-hydroxycinnamic acid is "fresh." We recommend recrystallizing this matrix from hot methanol and storing it in the dark at room temperature.
11. Use the matrix that is best for detection of the target peptide—e.g., 2,5-dihydroxybenzoic acid for phosphopeptides.
12. You may need to cut the tip of the pipette tip off with a clean razor blade or scalpel in order to efficiently transfer the beads.
13. One of these CRC columns will be used for the "control" beads. Control beads are useful for detecting nonspecific binding.
14. The manufacturer's *general* protocol says "1 h at room temperature or overnight at 4 °C," but in the section on factors controlling binding, they also state that binding is usually "completed" within 2 h at room temperature. This reference also includes several sections on how to adjust the coupling if you get too much or too little binding [11].
15. If you are using a valuable antibody, save the eluant! If, for some reason, the antibody doesn't bind to the beads, or if the amount of antibody used exceeds the binding capacity of the column, the unbound antibody will pass through the CRC.
16. Recently, our laboratory has conducted experiments using prototype Dynabeads® stored in a buffer without a detergent. The handling of these beads is more difficult as the beads adhere to tubes and pipette tips more without the presence of detergent, but the lack of detergent-related spectrum noise has resulted in improved sensitivity for the detection of low-abundance peptides. However, these beads are not commercially available at this time. We have included this section in the hopes that these detergent-free beads will soon become commercially available.
17. If you have several samples, this step can be done as a batch. In other words, a large amount of bead slurry can be incubated in a large volume of CHAPS for washing. If the washes are done in a large batch, then resuspend the beads in the original volume. For example, if 50 μl of bead slurry was washed in 2 ml

of PBS-CHAPS, then resuspend the beads in 50 μl of 1× PBS once washing has been completed.

18. For every 5 μl of original bead slurry, it is optimal to keep the final incubation volume to approximately 100–150 μl. Larger volumes tend to interfere with proper binding.

19. In our experiments, a "one-step incubation" containing the beads, antibody, digest, and labeled peptide has worked successfully. We highly recommend trying this approach; however, it is possible that experiments done in this manner with different antibodies and peptides may not yield the same results.

20. If protease inhibitors are present in the lysis buffer of the tissue, an acetone precipitation will have to be performed prior to digestion with trypsin.

21. For subsequent incubation with antibodies, a solid-phase extraction will have to be performed to remove reducing agents from sample. Our laboratory successfully uses Waters Oasis HLB columns according to manufacturer's protocol.

22. Add SIS peptides after digestion of the sample, not before.

23. If you are using a valuable lysate, save the eluant! If, for some reason, the antibody doesn't bind the peptides, or if the amount of peptide in the sample exceeds the binding capacity of the column, unbound peptides will pass through the antibody column. You can also save the eluent in case you later decide to study additional peptides from a different protein.

24. It should be noted that the peptides used for the EGFR versus EGFRvIII project both contain cysteine, which we usually try to avoid in iMALDI target peptides because of the possibility of the cysteine residue forming a disulfide bond. For this project, however, we were forced to select peptides that were different between these two variants, and these peptides happened to contain a cysteine. We did determine, however, that these peptides did not reoxidize in the samples, and could be captured and analyzed after reduction but without alkylation of the sample. As an additional complication, in this particular case, our standard alkylation methods (i.e., with DTT and iodoacetamide) would have shifted the molecular weight of the underivatized normal peptide by 57 Da, which would have made the derivatized normal peptide isobaric with the underivatized variant peptide, which contains an extra glycine residue. If the derivatization were not 100 % complete, this may have led to quantitation problems, although the two forms could be differentiated on the basis of the MS/MS spectra. If we were to develop this into a clinical assay, however, it would probably be better to use reduced and alkylated peptides to generate the antibody, and to reduce and alkylate the digest in the same way. Even this, however, would be somewhat challenging because of

the need for a free N-terminal cysteine to conjugate to the carrier protein to enhance its antigenicity during antibody production. We have learned, however, that it *is* possible to synthesize the target peptides with pyridylethylcysteine instead of the "internal" cysteines for the purposes of antibody generation. The free cysteines in the sample would then be modified by pyridylethylation before capture, and the capture would be done using an antibody raised against a target peptide which likewise contained pyridylethylcysteine. An alternative would be to develop protocol for reduction without alkylation, or for a reversible alkylation protocol such as MMTS. In any event, although this discussion is somewhat beyond the scope of this protocol, it serves to illustrate the types of considerations that must go into the planning of an iMALDI experiment.

Acknowledgements

The EGRFvIII work was supported by Genome BC, Genome Canada, and graduate student fellowships from the University of Victoria.

The *Francisella* study was funded by the Southeast Regional Center of Excellence for Emerging Infections and Biodefense (SERCEB, 5U54AI057157-04); the EGFR work was partially supported by NIH grant P30 CA16086-25, CEHS grant P30ES10126, SBIR grant HHSN261200433014C, and P50-CA58223-09A1. We would also like to acknowledge Jian Jiang and Erin Warren, who performed some of the work shown here.

References

1. Parker CE, Tomer KB (2000) Epitope mapping by a combination of epitope excision and MALDI-MS. Methods Mol Biol 146: 185–201
2. Jiang J, Parker CE, Fuller JR, Kawula TH, Borchers CH (2007) An immunoaffinity tandem mass spectrometry (iMALDI) assay for detection of *Francisella tularensis*. Anal Chim Acta 605:70–79
3. Jiang J, Parker CE, Hoadley KA, Perou CM, Boysen G, Borchers CH (2007) Development of an immuno tandem mass spectrometry (iMALDI) assay for EGFR diagnosis. Proteomics Clin Appl 1:1651–1659
4. Shah B, Borchers CH (2009) iMALDI: a targeted proteomics approach to the differentiation of EGFR and its isoforms. In: Presented at the 57th ASMS conference on mass spectrometry and allied topics, Philadelphia, PA, 31 May–4 June 2009
5. Bordeerat NK, Georgieva NI, Klapper DG, Collins LB, Cross TJ, Borchers CH, Swenberg JA, Boysen G (2009) Accurate quantitation of standard peptides used for quantitative proteomics. Proteomics 9:3939–3944
6. Warren MRE, Parker CE, Mocanu V, Klapper DG, Borchers CH (2005) Electrospray ionization tandem mass spectrometry of model peptides reveals diagnostic fragment ions for protein ubiquitination. Rapid Commun Mass Spectrom 19:429–437
7. Proc JL, Kuzyk MA, Hardie DB, Yang J, Smith DS, Jackson AM, Parker CE, Borchers CH (2010) A quantitative study of the effects of chaotropic agents, surfactants, and solvents on the digestion efficiency of human plasma proteins by trypsin. J Proteome Res 9: 5422–5437
8. Kuzyk M, Smith D, Yang J, Cross T, Jackson A, Hardie D, Anderson L, Borchers C (2009)

MRM-based, multiplexed, absolute quantitation of 45 proteins in human plasma. Mol Cell Proteomics 8:1860–1877

9. Sigma–Aldrich (2009) Peptide stability. http://www.sigmaaldrich.com/life-science/custom-oligos/custom-peptides/learning-center/peptide-stability.html. Accessed 15 Aug 2009
10. Parker CE, Warren MR, Loiselle DR, Dicheva NN, Scarlett CO, Borchers CH (2005) Identification of components of protein complexes. Methods Mol Biol 301:117–151
11. GE_Healthcare (2009) Instructions 71-7086-00 AF Affinity media, CNBr-activated Sepharose™ 4B. https://www.gelifesciences.com/gehcls_images/GELS/Related%20Content/Files/1314787424814/litdoc71708600AF_20110831144357.pdf. Accessed 1 June 2013

Chapter 7

Analysis of Neuropeptides by MALDI Imaging Mass Spectrometry

Anna Ljungdahl, Jörg Hanrieder, Jonas Bergquist, and Malin Andersson

Abstract

Matrix-assisted laser desorption ionization (MALDI) imaging mass spectrometry (IMS) is one of the most effective tools for localizing small molecules and compounds directly in thin tissue sections. MALDI IMS should be used when the distribution of molecular species is not known and to localize changes due to a disease process or a treatment. In recent years it has become increasingly clear that many pathological processes are not readily correlated to dramatic changes in protein levels. MALDI IMS can aid the localization of areas where the cellular concentration of proteins may be high enough to play an important biological role, but when the precise location is unknown. Here, we present a MALDI IMS protocol and data analysis of molecular imaging of multiple rat brain sections.

Key words MALDI, Matrix-assisted laser desorption/ionization, Imaging mass spectrometry, Neuropeptides, Reproducibility, Brain

1 Introduction

Matrix-assisted laser desorption ionization (MALDI) imaging mass spectrometry (IMS; [1, 2]) offers a unique analytical method to determine localization of molecular species in combination with high molecular specificity and sensitivity. The localization of neuropeptides and small proteins is particularly important in brain research considering that the brain is highly spatially and somatotopically organized.

MALDI IMS has been used for the analysis of many different kinds of molecules, including neuropeptides and proteins [1], phospholipids [3, 4], trypsinized proteins [5], lipids [6], drugs [7–9], amyloid plaques [10], etc. (for protocols *see* [11]). The analysis is typically focused on different molecular species by the choice of matrix or chemical pretreatment, but it is also possible to combine the MALDI source with different mass analyzers, for

Helena Bäckvall and Janne Lehtiö (eds.), *The Low Molecular Weight Proteome: Methods and Protocols*, Methods in Molecular Biology, vol. 1023, DOI 10.1007/978-1-4614-7209-4_7,

example, in order to increase mass accuracy one can use Fourier transform ion cyclotron resonance MS (FTICR MS; [12]) or to resolve lipids from proteins ion mobility separation can be used [13]. In this chapter we focus on the practical aspects of sample preparation of multiple samples, MALDI matrix application, MALDI-TOF MS acquisition, and data analysis of neuropeptides and small brain proteins in fresh frozen tissue sections.

2 Materials

2.1 Sample Collection and Preparation

1. Isoflurane (Apoteksbolaget AB, Sweden).
2. Surgical tools; two pairs of scissors, a rongeur, spatula.
3. Finely crushed dry ice.
4. Aluminum foil and plastic bags for storage.
5. Cryostat (Microm, Heidelberg, Germany).
6. Indium tin oxide (ITO)-coated MALDI slides (Bruker Daltonics, Germany).
7. Optimal cutting temperature medium (OCT) (Sakura Finetek Europe, The Netherlands).
8. Vacuum desiccator.

2.2 Matrix Application

1. Acetonitrile (ACN), methanol (MeOH), ethanol (EtOH) of pro-analysis.
2. Trifluoroacetic acid (TFA).
3. Canister of compressed air.
4. Water was purified with a Milli-Q purification system (Millipore, Bedford, MA).
5. CHIP-1000 (Shimadzu, Japan) or Portrait 630 spotter (LabCyte, Sunnyvale, CA).

2.3 Mass Spectrometry

1. Ultraflex II (Bruker Daltonics).
2. FlexImaging software (v 2.0, Bruker Daltonics).
3. Peptide or Protein Calibration Standard (Bruker Daltonics).

2.4 Data Processing

1. Origin v.8.1 (OriginLab Corp, Northampton, MA).
2. Matlab version 7.9.0.529 (MathWorks, Natick, MA).
3. pBin (http://www.vicc.org/biostatistics/software.php).

2.5 Neuropeptide Identification

1. Ultrasound sonicator.
2. 10 kDa molecular weight cut-off filter (Millipore, Bedford, MA).
3. SP Sephadex C 25 gel (Sigma, Germany).

4. Hypersil column (ThermoFisher).
5. Pyridine, formic acid (FA), ACN, MeOH, EtOH of pro-analysis grade.
6. TFA.
7. Agilent 1100 nanoflow LC, microWPS, C18 column (0.075 × 150 mm).
8. Agilent 1100 Micro-fraction collector, Prespotted anchorchip target plate (Bruker Daltonics).
9. BioTools software (v 3.1 SR2, Bruker Daltonics).
10. Bio Works software (v.3.3, Thermo Fisher; www.thermoscientific.com).
11. Mascot software (v 2.2, MatrixScience; www.matrixscience.com).
12. Swepep database (www.swepep.org).

3 Methods

3.1 Sample Collection and Preparation

1. For MALDI IMS of peptides and small proteins from fresh frozen tissue it is important to work fast and diligent during dissection (*see* **Note 1**). Protein degradation will start immediately (*see* **Note 2**). Anesthetize animal with isoflurane. The time from decapitation to placing the dissected tissue on finely crushed dry ice should be less than 30 s. Cover tissue entirely with powdered dry ice and freeze for about 5 min before wrapping in aluminum foil and transfer to a container (plastic bag) for long-term storage at −80 °C. Do not freeze tissue directly in test tubes since it is important to preserve the tissue's original structure and form.
2. Cut sections on a cryostat and thaw mount onto a MALDI target (*see* **Notes 3** and **4**; Fig. 1). Avoid contamination of sections with embedding media; these are commonly based on polymer mixtures including polyethylene glycol that will impair the mass spectrometry signal (e.g., Tissue Tek, OCT media). A contaminated knife blade should be discarded. Freeze damages can occur during sectioning and are often clearly visible as white icy areas (Fig. 1b–d). Discard sections with microtears and cracks, as they will severely impair matrix crystallization and thus result in poor MS quality (*see* **Note 5**).
3. In order to prevent protein degradation and freeze damage, drying the sections after thaw-mounting is pertinent. The best protocol includes a rapid transfer of a mounted section directly to a vacuum desiccator for 15 min of drying and then to proceed to apply the matrix. However, this is not practical for any large-scale MALDI IMS analysis that includes several experimental

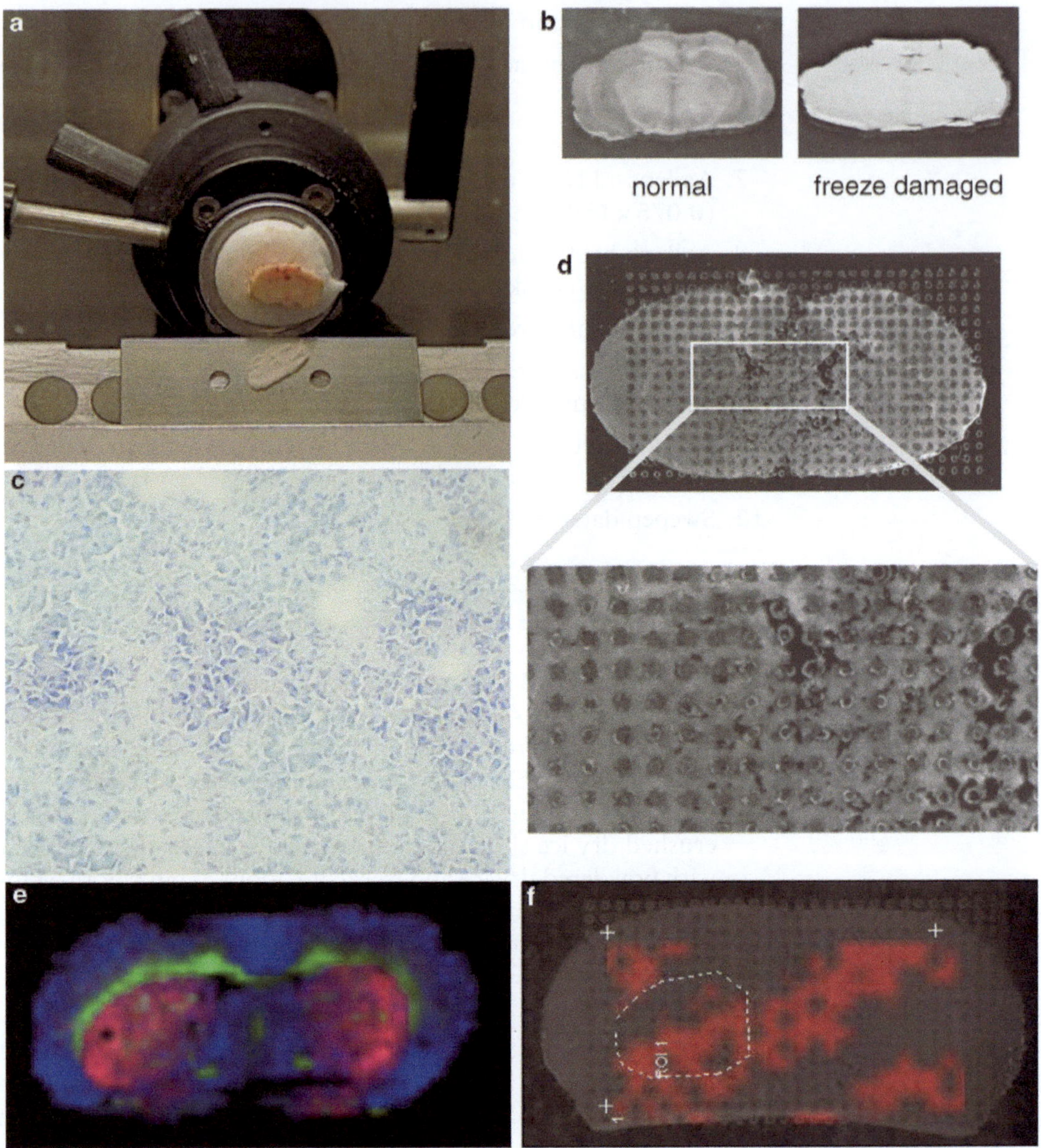

Fig. 1 **(a)** The brain is mounted on a cryostat chuck using an embedding media (OCT). Cryostat sections are cut at 12 μm thickness. (**b**) The sections are thaw-mounted onto MALDI-compatible glass slides and dried for a few seconds to avoid freeze damage. (**c**) A section stained with *cresyl violet* reveals microtears and cracks. (**d**) Microtears may be difficult to detect with the naked eye, but impair MALDI matrix crystallization and obliterate the MALDI MS signal. (**e**) Three different proteins are visualized in a Fleximaging (myelin basic protein in *green*, ubiquitin in *blue*, and PEP-19 in *red*). (**f**) Misalignment of MALDI-TOF motor coordinates to image scan can cause a systematic shift that renders an IMS experiment useless

groups and samples. We find that 5-s drying time as the section is thaw-mounted will protect sufficiently from protein degradation and allow the mounting of several sections onto the same MALDI target. Once the target is full of sections they can be transferred to the vacuum desiccator for a 15–20-min drying.

4. Dried sections can be stored at −20 °C for a few weeks or at −80 °C for a few months without any noticeable change in MALDI MS signal. To prevent oxidation of peptides and proteins the slides should be placed in a container flushed with inert gas (e.g., nitrogen).
5. Defrost sections by a first transfer from −80 to −20 °C for a few hours and then thaw in a vacuum desiccator for at least 30 min.

3.2 Matrix Application

1. Wash tissue (*see* **Note 6**). Wash all slides in a cuvette at the same time so that they are all treated in exactly the same way. For protein MALDI IMS the washing can be rather extensive; submerse MALDI target in 70 % ethanol for 30 s, followed by two washes in 95 % ethanol for 30 s. For peptide MALDI IMS the washes have to be short and precisely timed, as peptides are typically water soluble and will be lost with extensive washing. We submerge brain sections in 70 % ethanol for 10 s, followed by two submersions in 95 % ethanol for 10 s each. Dry face up in vacuum desiccator.
2. Choose MALDI matrix, for example DHB for peptides or sinapinic acid for proteins (*see* **Note 7**).
3. Seeding is a term used to apply a thin layer of MALDI matrix crystals onto tissue prior to applying a MALDI matrix solution. Seeding can improve MS signalling and spot precision. Use a mortar and pestle to grind the MALDI matrix to a fine powder. Use a brush, spatula, or sieve to deposit the powder evenly on top of section. Remove excess powder by a jet stream of compressed air.
4. Mix the MALDI matrix with solvent (20 mg/ml DHB in 50 % MeOH, 0.3 % TFA in water, or 25 mg/ml sinapinic acid in 50 % ACN, 0.3 % TFA in water). Test by applying 0.2 μl-sized drops of matrix directly on tissue using a pipette. Hand-spotting often results in good mass spectra quality; however the reproducibility is often very low with a %CV of around 50 [27].
5. Choose MALDI matrix deposition strategy (*see* **Note 8**). We mainly use *automatic drop deposition* by a chemical printer or an acoustic ejector (e.g., ChIP100, Shimadzu or Portrait630, LabCyte), but dry coating (*see* **Note 9**) and spray coating (*see* **Note 10**) is also commonly used.
6. Deposit one or several 90–200 pl-sized drops at each location, creating a "spot" of matrix mixed with analyte. Determine the optimum ratio between the amount matrix and analyte by varying the number of drops per spot and/or the number of spot cycles (*see* **Note 11**). It is important that the time it takes to print a whole array allows for both protein/peptide extraction into the matrix solvent, and evaporation of solvent with resulting co-crystallization of the matrix and analyte (typically 2 min).

If there is not enough time for drying between deposits, the added volume of matrix may cause neighboring spots to merge and result in delocalized peptides.

7. Assess the optimum protocol on the level of mass spectrum quality, in terms of background, noise level, base peak signal intensity, peak resolution, and number of peaks detected (*see* **Note 12**).
8. As for any proteomic experiment or traditional biochemical analysis, reproducibility is the key for a successful analysis. Run as many samples as possible in one single session. Include proper controls, such as standard peptide/protein mix for mass accuracy and a series of dilutions for test of MALDI signal sensitivity. Blind the experiment and relabel slides.

3.3 Mass Spectrometry

1. Use one of a number of publicly or commercially available MALDI IMS acquisition software that can be used with different MALDI mass spectrometers. We use FlexImaging and although acquisition procedures vary with each software and mass spectrometer, some aspects are common for all.
2. *Registration*. Scan the matrix-covered tissue sections in a flatbed scanner. For sections with printed arrays it is often easy to find good registration marks; however, even small shifts in inaccurate registration can affect the MALDI signal quality and in the worst case render the experiment worthless (Fig. 1e, f).
3. *MALDI acquisition method*. Determine the optimum number of good-quality spectra that can be obtained from each position. It is important to avoid over-sampling and summing up noise alone. For a typical DHB matrix spot we sum up 500 laser shots from 20 locations within each spot, and for sinapinic acid matrix we sum 200 shots from ten locations. Use a movement raster within (sinapinic acid) or on the edges (DHB) of each spot (*see* **Note 13**).
4. *Calibration*. Always include at least two calibration spots placed in two corners diametrically over slide, and save the spectra so that you can go back later and recalibrate! If the signal is not as good as expected, double-check the calibration spectra with earlier experiments and if doubting the MALDI sensitivity use a BSA digest dilution series to determine sensitivity: 400, 40, 4, and 0.4 fmol/μl. Take 0.25 μl digest and mix on target with 0.25 μl matrix to get 100, 10, 1, and 0.1 fmol, respectively, on target.
5. *MS processing*. Add on-the-fly MS processing and export as ascii files, using a baseline removal algorithm that will not result in negative data (e.g., ConvexHull 0.8 flatness; FlexAnalysis, Bruker). In this system, the raw data is preserved and changing a processing is always possible.

6. Before changing to the next MALDI target, double-check that all sections on the first target have been run completely in the MALDI IMS image viewer software (FlexImaging, Bruker; Biomap, Novartis, Switzerland; *see* **Note 14**).

3.4 Data Processing

1. Two approaches are typically used for data analysis of MALDI IMS experiments. One is based on the digital information contained in the images for each peak and the other utilizes common MS-based tools to perform recalibration, baseline subtraction, noise reduction, and normalization of each mass spectrum in the analysis (*see* **Note 15**). In our lab, we export the mass spectra from areas of interest and perform MS analysis separately from visualization.
2. *Normalization*. For each mass spectra, sum up every intensity measurement to get the total ion current (TIC; *see* **Note 16**). Assess whether TIC normalization should be employed by comparing the TIC (before normalization!) for the test vs. control groups; if the TIC is significantly different there is a high risk of false positives because of over-normalization effects in one group versus the other. Visualize the TIC in FlexImaging or Biomap (www.maldi-msi.org) to reveal areas of tissue with low TIC values, typically induced by tissue damage or freeze artifacts. If TIC normalization can be employed, then divide each intensity measurement by its respective TIC and multiply with a scaling factor (typically 10^5 or 10^6, equal to the average order of magnitude of all TIC in the experiment).
3. *MS recalibration*. If several MALDI target holders are used, a slight shift in calibration accuracy can occur. Use the algorithm *msalign* available in the Matlab bioinformatics tool pack to realign the spectra. Use a subset of peaks that are common to most mass spectra as calibration markers (Fig. 2).
4. *Peak detection*. The amount of data produced in a typical MALDI IMS experiment can easily overwhelm the most experienced scientists (*see* **Note 17**). Use a peak picking algorithm based on detecting local maxima and a signal-to-noise threshold of >3 to select peaks (Origin or MatLab). Typically we do peak picking on each mass spectrum individually in order to obtain high peak detection sensitivity (*see* **Note 18**).
5. *Peak binning*. The result of peak picking is a series of peak lists where each peak apex is followed by the peak intensity at that data point. Binning peaks set all peaks on a common m/z scale and it is a simple way to make data manageable for statistical analysis. We use a standalone script, p-Bin (http://www.vicc.org/biostatistics/software.php), to generate bins with a start mass and an end mass (Fig. 2c). For pBin, select a minimum number of spectra that contain a specific peak (expressed as % of total nb mass spectra) and a linear function describing the width of the peaks throughout the m/z axis. This typically has

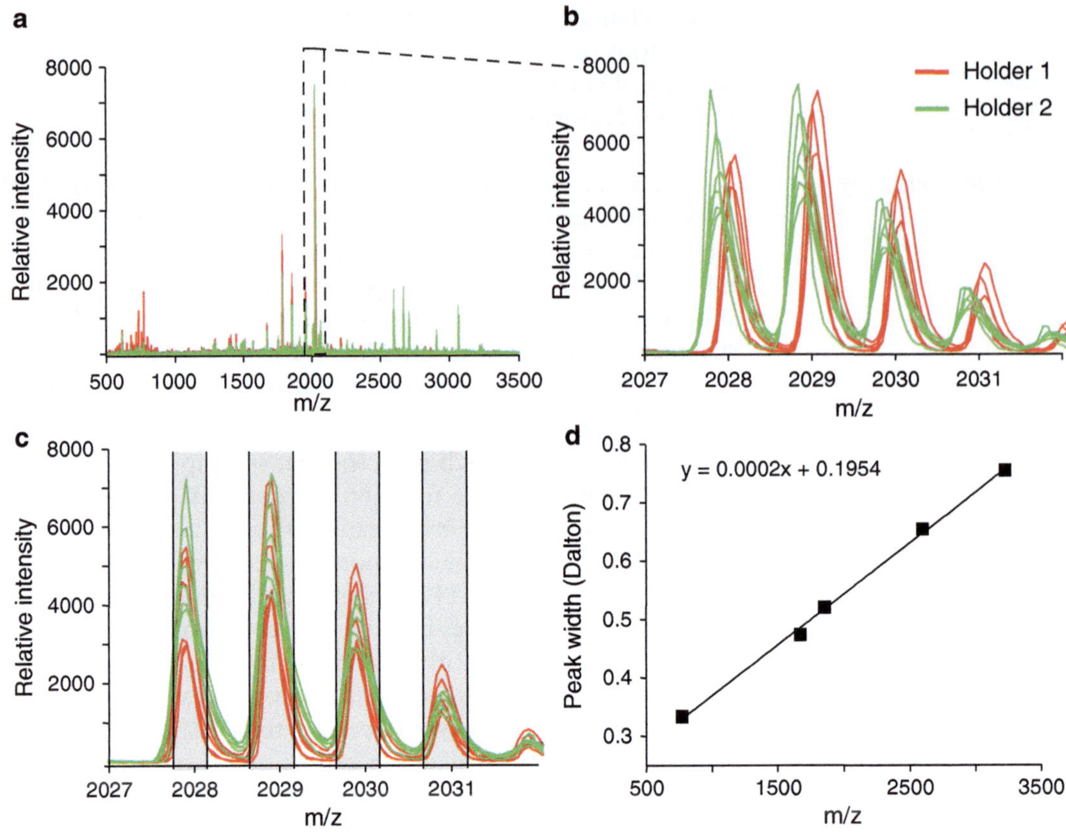

Fig. 2 **(a)** IMS of brain sections from two different MALDI target holders (*red* traces from holder 1 and *green* traces from holder 2). (**b**) The magnified view reveals a 0.1 Da small mass shift between the two holders. (**c**) Realignment of mass spectra will greatly improve peak picking and binning of peaks. The peak area is calculated within the limits of each bin (*shaded area*). (**d**) For binning it is important to measure the peak width and calculate a linear function for the mass range analyzed

to be adjusted for every experiment, for example; for small proteins $y = 0.0012X + 2.98$; for peptides $y = 0.0002X + 0.195$ (Fig. 2d). One output file contains the peak intensities sorted into a matrix that can be used directly for statistical analysis. These peak intensities represent the peak height of peaks with a signal-to-noise ratio > 3, present in X% of the total number of spectra. However, peak areas can provide a more accurate measurement and for this purpose the pBin output file "binrange" can be used to integrate the area under the curve (AUC) for each peak in MatLab.

6. *Data evaluation.* Calculate average peak AUC for each brain section and compare the overall variation within groups by calculating the relative standard deviation (RSD = SD/peak AUC) for all peaks detected. Plot the RSD as a function of m/z and determine if the within-group variation overlaps (average RSD

ranging from 16 to 22 %; Fig. 3a) or not (Fig. 3b). In the example presented in Fig. 3b, the overall RSD is extremely high, but more importantly a significantly higher variance was detected by one-way ANOVA in the control (45 %) and group A (41 %) vs. group B (37 %). To determine if single sections are the source of variation, plot all peak intensities in pairs of mass spectra from animals within the same group (Fig. 3c, d). Exclusion of mass spectra that do not display linearity (Fig. 3d) can improve overall variation; in example 3B the control group variation improved from 45 to 29 % and in group A from 41 to 31 %.

7. *Statistical analysis.* The data can be evaluated with many different statistical analyses, including *t*-tests, parametric or non-parametric tests of variance, the Excel plug-in SAM (Stanford Statistical Analysis of Microarray data; [14]), LIMMA [15], or principal component analysis (PCA).

3.5 Neuropeptide Identification

1. Sometimes the identity of neuropeptides with a mass up to ~3,000 Da can be verified directly from a spotted tissue section by tandem mass spectrometry [16].
2. Peptide enrichment is often necessary (*see* **Note 19**). Peptide extraction can be performed using (a) 5 % methanol and 0.1 % TFA applied directly onto tissue sections or (b) with 1 M acetic acid for tissue pieces. Homogenize tissue pieces with ultrasonication and heat at 95 °C for 5 min. Centrifuge for 20 min at 9,000 × *g*, and collect supernatant. Filter supernatant through a 10 kDa molecular weight cutoff spin filter.
3. An additional separation step can reduce sample complexity using an ion exchange column packed with SP Sephadex C 25 gel. Elute with increasing amounts of pyridine (in 0.1 % formic acid) stepwise with 0, 30, 70, and 100 % pyridine. Dry fractions in a vacuum centrifuge, reconstitute the IEX fractions in 50 μl 0.1 % formic acid, and store at −20 °C.
4. Set up an HPLC system capable of a nanoscale flow rate and prepare fresh buffers A: 0.1 % formic acid and B: 100 % ACN, and 0.1 % formic acid. Inject 5 μl sample onto a C18 column using a manual six-port valve or an autosampler and run a gradient starting from 2 % B for 10 min, then from 2 % B to 50 % B in 40 min and afterwards from 50 % B to 95 % B in 5 min, maintain 95 % for 5 min for column washing, and finally back to 2 % B for reconstituting the column. In case of LC-MALDI, peptide elution is followed by offline fractionation using a micro fraction collector capable for direct spotting on MALDI target plates.
5. MALDI analysis of manually extracted peptides is performed with an Ultraflex II instrument (Bruker) running in reflector positive mode. Candidate masses of interest are subjected to MSMS analysis.

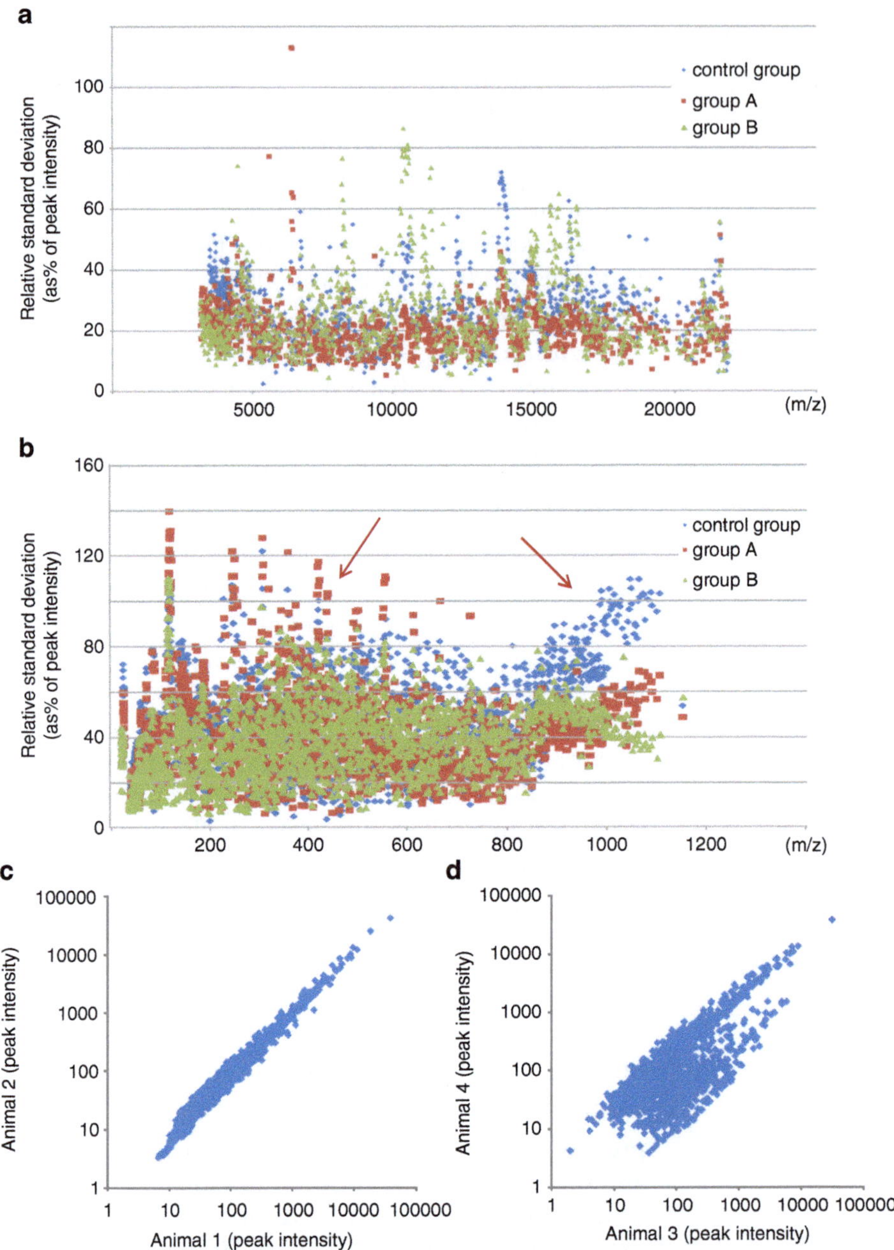

Fig. 3 **(a)** The relative standard deviation (RSD), expressed as standard deviation/peak intensity, can be used as a simple tool to analyze overall experimental success. High-intensity peaks often display higher standard deviation, but the mean RSD ranges from 16 to 22 % in each group ($n=6$ rats per group). (**b**) A generally high RSD indicates that the IMS experiment was unsuccessful, but in addition, the control group and group A display significantly higher RSD compared with group B (*arrows*). (**c**) Intensity–intensity plots of peaks from animals in the same group display equal peak intensities, close to a 1:1 ratio, indicating good MS quality in spectra from both animals. (**d**) Intensity–intensity plots from animals where the peak intensities do not correlate linearly, revealing that one or both animals are defective. In order to exclude one or the other, the next step is to plot and compare animals 3 and 4, separately, against the peak intensities of animal 1

6. For LC-ESI FTICR experiments, the nanoflow LC is hyphenated directly to the electrospray ion source and peptide elution is followed online by MS and MSMS analysis. A full-scan spectrum is acquired at high resolution using the FT analyzer. Data-dependant acquisition is applied for MSMS precursor selection, where the five most intense mass peaks are subjected to subsequent isolation and collision-induced fragmentation in the ion trap. The MSMS fragment ions are detected using the LTQ or the FT analyzer.
7. The annotated fragment spectra are subjected to database search using a search engine software of choice (i.e., Mascot, Sequest, X!Tandem, Proteinpilot, MS-tag). For peptide identification database containing known neuropeptide sequences is used together with a database containing sequences of known neuropeptide precursor proteins (Swepep, [17]).

4 Notes

1. The heme group of hemoglobin (*m/z* 616) is easily ionized and desorbed in the MALDI process and can dominate a spectrum through ion suppression, whereby signals of other proteins and peptides are suppressed despite being quite abundant. Dissecting quickly will minimize the collection and coagulation of blood in blood vessels throughout the tissue. An alternative is to flush blood vessels with buffer, for example using saline for transcardial perfusion. Although this is highly effective for removing blood, this can increase collection time and cause increased protein degradation. Several other sources of tissue may be considered, including formaldehyde- [18] and ethanol- [19] fixed tissue.
2. Other alternatives to combat protein degradation are the use of focused microwave preservation or to denature the proteins in the dissected tissue through rapid heating [20–23]. Rapid freezing in liquid nitrogen often causes cracks and tears in tissue that interfere with MALDI IMS.
3. Choosing the right tissue section thickness will affect several parameters downstream, but mainly it will affect the amount of MALDI matrix needed to obtain the optimal MS quality. Thin sections (3–6 μm) will require much less MALDI matrix, which will in turn speed up the matrix application process. On the other hand, thin sections of fresh frozen tissue are notoriously more difficult to cut on a cryostat or a freezing microtome. In general, 12 μm thick sections are easy to manipulate and thaw-mount onto a MALDI target.

4. The choice of MALDI target depends mainly on the MALDI-TOF mass spectrometer used and generally the commercially available targets are acceptable. Different kinds of MALDI targets are available: ITO-coated glass slides, stainless steel, gold coated, etc. The choice of target may influence desorption and ionization of ions, but it is generally possible to adapt the matrix protocol to obtain good peptide and protein MALDI IMS. We prefer the ITO-coated glass slides since they are compatible with histological staining and subsequent microscopic survey of the tissue after MALDI analysis.
5. Microscopic inspection of the target reveals that brain tissue integrity is preserved for about 36–48 h in the vacuum desiccator, after which tissue distortion, microtears, and small cracks will appear and MALDI IMS can be compromised. The time until deterioration is dependent on the type and thickness of the tissue stored.
6. Washing tissue serves two main purposes: first to remove salts and debris that interfere with the MALDI MS process, and second to fixate proteins and protect against protein degradation during the time it takes to apply the matrix. Several alcohol-based washing protocols have been described that enhance MALDI IMS quality [16, 24, 25].
7. The typical matrix is a small organic acid that absorbs energy around 355 nm, sublimating together with analyte and transferring charge to the analyte. The matrix is typically dissolved (at 10 mg/ml) in acidified part organic and part aqueous solvent, e.g., 0.1–0.5 % TFA plus 50 % ACN in water, or 0.3 % TFA plus 50 % methanol in water. The TFA lowers the pH of the matrix to around pH 2, which ensures that most peptides and proteins will be easily protonated and ionized. The most commonly applied MALDI matrices for proteins are sinapinic acid (SA) and 2,5-dihydroxyacetophenone (DHAP). Common MALDI matrices for peptides and tryptic peptides are α-cyano-4-hydroxycinnamic acid (CHCA) and dihydroxy-benzoic acid (DHB).
8. When it comes to matrix deposition it appears to be most difficult to have it all: high imaging resolution, great MALDI MS signal, excellent data reproducibility, and speed. For high resolution that is only limited by the size of the laser beam and/or size of MALDI crystals, dry-coating (similar to seeding, Subheading 3.2, **step 1**) or spray-coating the MALDI matrix is a good option. For high sensitivity and reproducibility automatic drop deposition can be recommended.
9. Dry-coating the MALDI matrix without any solvent is especially effective for detection of lipids [4]. Grind matrix with mortar and pestle, and sieve to achieve a specified particle size (e.g., <40 μm). Use a sieve (about 100 μm) to deposit the powder evenly on top

of section. Remove excess powder. Repeat four to five times until the section is completely covered.

10. Use a pneumatic sprayer or an automatic MALDI matrix sprayer (e.g., ImagePrep, Bruker Daltonics, or TM sprayer, Leap Technology). Determine the number of spray cycles that are needed for an optimum ratio of matrix and analyte. Avoid flooding the tissue section as this causes the peptides and proteins to delocalize. Using automatic sprayers it is possible to repeat the matrix deposition exactly the same way every time; however any matrix deposition method that relies on a spray will produce a Gaussian distribution of different size of droplets with the highest concentration of drops and largest droplets in the center of the spray. This can cause differences in the MALDI MS signal from center to periphery of the target, some of which can be prevented by rotating the target between spray cycles and some corrected by later MS normalization (see below).

11. Automatic drop deposition permits highly efficient protein/peptide extraction without risking molecular delocalization across the tissue [26]. A too low or too high total amount of matrix will result in poor signal-to-noise ratio as well as sensitivity. Working with very high concentrations of matrix may crystallize upon contact with air and will compromise the printer/spotter function, and most protocols use lower concentrations of matrix (5–10 mg/ml) and revisit the same spot over and over again in order to deposit large amounts of matrix.

12. Deposition of a high number of matrix solution drops per pass and passes may result in great MS signal quality. However, bigger spots will reduce the maximum image resolution that can be achieved, and furthermore, we find that applying high volumes (2–2.5 nl) of matrix each pass can affect the print position precision severely if the volume deposited exceeds the surface tension, the drop tilts and moves slightly on the tissue and thus result in poor MALDI imaging MS quality. This can be corrected if every matrix spot position is given individual picture/motor coordinates for MS acquisition.

13. DHB matrix crystallizes around the edges of drop deposition and thus results in a ringlike structure that displays different molecular MS signatures from the edge versus the middle. Most peptide peaks will be detected around the edges and thus correct acquisition will focus on sampling the edges more than the center.

14. Common pitfalls include mapping the maximum absolute peak intensity to a linear 100 % optical scale for every section in the experimental series, instead of mapping all sections to a common optical density scale where a 100 % equals the maximum peak intensity of *all* sections. It is often possible to screen for

differences within the same section or imaging experiment by selecting regions of interest (ROI) and plotting average mass spectra for comparison of different areas. This method can predict large differences but is often not conclusive, since comparisons typically include several sections and experimental conditions. It is our experience that data analysis ought to be done on the spectral data and not on imaging data or a single average spectrum.

15. Many different data processing and visualization tools are commercially available, e.g., TissueView (Applied Biosystems), ImageQuest (ThermoScientific), and MassLynx (Waters Inc.). For image baseline and noise reduction, *see* [27]. For MS-based processing of IMS data *see* [28]. Mass spectra overall quality and reproducibility is a central issue in protein analyses [29, 30]. Experimental sources of variation in MALDI IMS are similar to other MS platforms, such as MALDI-TOF and SELDI-TOF profiling MS, and depend on many factors, including tissue/sample preparation, matrix deposition, MS acquisition, and MS processing [31, 32].

16. TIC normalization is a commonly used strategy, straightforward in form and easily implemented for different applications [28, 33–35]. The underlying basis for normalization is to remove the effects of systematic errors related to a constant factor, mainly sampling efficacy, and that the sum intensity in each spectrum should be equal given that the underlying tissue contains equal amounts of protein. By contrast to normalizing against a base peak, the sum of peak heights, or peak areas, the TIC relies less on changes in individual peak intensities due to biological factors and more on other sources of variation that affect every ion equally throughout the mass range.

17. For example, take the smallest experiment focusing on one organ from two groups of subjects, where $n = 3$ and 3 in control and treated group, respectively. Take one section from each organ, and image using a relatively low resolution summing up 100×100 pixels (or spectra) over a mass range of 2–50,000 Da recorded in 50,000 data points (m/z and intensity). In total this will add up to 60,000 spectra, each containing two columns of 50,000 data points, yielding an overall matrix of 6×10^9 numbers to calculate. Thus data reduction is of essential interest.

18. An alternative is to create one average mass spectrum for each organ/test subject and this often results in lower noise levels but also lower peak resolution of small peaks, essentially drowning them in noise. The advantage of doing peak picking on each mass spectrum is that the increased sensitivity may reveal localization patterns when visualized, permitting true image analysis capability.

19. A main obstacle in peptidomics is the identification process, since common proteomic strategies are based on tryptic peptides and cannot be easily applied to endogenous peptides. Since tissue analysis by accurate mass matching is not sufficient for high-confidence identification further off-line strategies have to be used that allow significant sequence assignment.

References

1. Caprioli RM, Farmer TB, Gile J (1997) Molecular imaging of biological samples: localization of peptides and proteins using MALDI-TOF MS. Anal Chem 69:4751–4760
2. Stoeckli M, Chaurand P, Hallahan DE, Caprioli RM (2001) Imaging mass spectrometry: a new technology for the analysis of protein expression in mammalian tissues. Nat Med 7:493–496
3. Burnum KE, Cornett DS, Puolitaival SM, Milne SB, Myers DS, Tranguch S, Brown HA, Dey SK, Caprioli RM (2009) Spatial and temporal alterations of phospholipids determined by mass spectrometry during mouse embryo implantation. J Lipid Res 50:2290–2298
4. Puolitaival SM, Burnum KE, Cornett DS, Caprioli RM (2008) Solvent-free matrix dry-coating for MALDI imaging of phospholipids. J Am Soc Mass Spectrom 19:882–886
5. Groseclose MR, Andersson M, Hardesty WM, Caprioli RM (2007) Identification of proteins directly from tissue: in situ tryptic digestions coupled with imaging mass spectrometry. J Mass Spectrom 42:254–262
6. Jackson SN, Woods AS (2009) Direct profiling of tissue lipids by MALDI-TOFMS. J Chromatogr B Analyt Technol Biomed Life Sci 877:2822–2829
7. Khatib-Shahidi S, Andersson M, Herman JL, Gillespie TA, Caprioli RM (2006) Direct molecular analysis of whole-body animal tissue sections by imaging MALDI mass spectrometry. Anal Chem 78:6448–6456
8. Reyzer ML, Caprioli RM (2007) MALDI-MS-based imaging of small molecules and proteins in tissues. Curr Opin Chem Biol 11:29–35
9. Reyzer ML, Hsieh YS, Ng K, Korfmacher WA, Caprioli RM (2003) Direct analysis of drug candidates in tissue by matrix-assisted laser desorption/ionization mass spectrometry. J Mass Spectrom 38:1081–1092
10. Stoeckli M, Knochenmuss R, McCombie G, Mueller D, Rohner T, Staab D, Wiederhold KH (2006) MALDI MS imaging of amyloid. Methods Enzymol 412:94–106
11. Rubakhin SS, Sweedler JV (2010) A mass spectrometry primer for mass spectrometry imaging. Methods Mol Biol 656:21–49
12. Cornett DS, Frappier SL, Caprioli RM (2008) MALDI-FTICR imaging mass spectrometry of drugs and metabolites in tissue. Anal Chem 80:5648–5653
13. McLean JA, Ridenour WB, Caprioli RM (2007) Profiling and imaging of tissues by imaging ion mobility-mass spectrometry. J Mass Spectrom 42:1099–1105
14. Efron B, Tibshirani R (2002) Empirical bayes methods and false discovery rates for microarrays. Genet Epidemiol 23:70–86
15. Smyth GK, Michaud J, Scott HS (2005) Use of within-array replicate spots for assessing differential expression in microarray experiments. Bioinformatics 21:2067–2075
16. Andersson M, Groseclose MR, Deutch AY, Caprioli RM (2008) Imaging mass spectrometry of proteins and peptides: 3D volume reconstruction. Nat Methods 5:101–108
17. Falth M, Skold K, Norrman M, Svensson M, Fenyo D, Andren PE (2006) SwePep, a database designed for endogenous peptides and mass spectrometry. Mol Cell Proteomics 5:998–1005
18. Groseclose MR, Massion PP, Chaurand P, Caprioli RM (2008) High-throughput proteomic analysis of formalin-fixed paraffin-embedded tissue microarrays using MALDI imaging mass spectrometry. Proteomics 8:3715–3724
19. Chaurand P, Latham JC, Lane KB, Mobley JA, Polosukhin VV, Wirth PS, Nanney LB, Caprioli RM (2008) Imaging mass spectrometry of intact proteins from alcohol-preserved tissue specimens: bypassing formalin fixation. J Proteome Res 7:3543–3555
20. Goodwin RJ, Dungworth JC, Cobb SR, Pitt AR (2008) Time-dependent evolution of tissue markers by MALDI-MS imaging. Proteomics 8:3801–3808
21. Goodwin RJ, Pennington SR, Pitt AR (2008) Protein and peptides in pictures: imaging with

MALDI mass spectrometry. Proteomics 8:3785–3800

22. Scholz B, Skold K, Kultima K, Fernandez C, Waldemarson S, Savitski MM, Söderquist M, Boren M, Stella R, Andren PE, Zubarev R, James P (2011) Impact of temperature dependent sampling procedures in proteomics and peptidomics—a characterization of the liver and pancreas post mortem degradome. Mol Cell Proteomics 10:M900229MCP200
23. Svensson M, Boren M, Skold K, Falth M, Sjogren B, Andersson M, Svenningsson P, Andren PE (2009) Heat stabilization of the tissue proteome: a new technology for improved proteomics. J Proteome Res 8:974–981
24. Schwartz SA, Reyzer ML, Caprioli RM (2003) Direct tissue analysis using matrix-assisted laser desorption/ionization mass spectrometry: practical aspects of sample preparation. J Mass Spectrom 38:699–708
25. Seeley EH, Oppenheimer SR, Mi D, Chaurand P, Caprioli RM (2008) Enhancement of protein sensitivity for MALDI imaging mass spectrometry after chemical treatment of tissue sections. J Am Soc Mass Spectrom 19:1069–1077
26. Aerni HR, Cornett DS, Caprioli RM (2006) Automated acoustic matrix deposition for MALDI sample preparation. Anal Chem 78:827–834
27. Klinkert I, McDonnell LA, Luxembourg SL, Altelaar AF, Amstalden ER, Piersma SR, Heeren RM (2007) Tools and strategies for visualization of large image data sets in high-resolution imaging mass spectrometry. Rev Sci Instrum 78:053716
28. Norris JL, Cornett DS, Mobley JA, Andersson M, Seeley EH, Chaurand P, Caprioli RM (2007) Processing MALDI mass spectra to improve mass spectral direct tissue analysis. Int J Mass Spectrom 260:212–221
29. Albrethsen J (2007) Reproducibility in protein profiling by MALDI-TOF mass spectrometry. Clin Chem 53:852–858
30. Hilario M, Kalousis A, Pellegrini C, Muller M (2006) Processing and classification of protein mass spectra. Mass Spectrom Rev 25:409–449
31. Dijkstra M, Vonk RJ, Jansen RC (2007) SELDI-TOF mass spectra: a view on sources of variation. J Chromatogr B Analyt Technol Biomed Life Sci 847:12–23
32. Villanueva J, Philip J, Chaparro CA, Li Y, Toledo-Crow R, DeNoyer L, Fleisher M, Robbins RJ, Tempst P (2005) Correcting common errors in identifying cancer-specific serum peptide signatures. J Proteome Res 4:1060–1072
33. Baggerly KA, Morris JS, Wang J, Gold D, Xiao LC, Coombes KR (2003) A comprehensive approach to the analysis of matrix-assisted laser desorption/ionization-time of flight proteomics spectra from serum samples. Proteomics 3:1667–1672
34. Dekker LJ, Dalebout JC, Siccama I, Jenster G, Sillevis Smitt PA, Luider TM (2005) A new method to analyze matrix-assisted laser desorption/ionization time-of-flight peptide profiling mass spectra. Rapid Commun Mass Spectrom 19:865–870
35. Zhang J, Zenobi R (2004) Matrix-dependent cationization in MALDI mass spectrometry. J Mass Spectrom 39:808–816

Chapter 8

Highly Multiplexed Antibody Suspension Bead Arrays for Plasma Protein Profiling

Kimi Drobin, Peter Nilsson, and Jochen M. Schwenk

Abstract

Alongside the increasing availability of affinity reagents, antibody microarrays have become a powerful tool to screen for target proteins in complex samples. Applying directly labeled samples onto arrays instead of using sandwich assays offers an approach to facilitate a systematic, high-throughput, and flexible exploration of protein profiles in body fluids such as serum or plasma. As an alternative to planar arrays, a system based on color-coded beads for the creation of antibody arrays in suspension has become available to offer a microtiter plate-based option for screening larger number of samples with variable sets of capture reagents. A procedure was established for analyzing biotinylated samples without the necessity to remove excess labeling substance. We have shown that this assay system allows detecting proteins down into lower pico-molar and higher pg/ml levels with dynamic ranges over three orders of magnitude. Presently, this workflow enables the profiling of 384 samples for up to 384 proteins per assay.

Key words Suspension bead array, Antibody array, Serum, Plasma, Labeling

1 Introduction

The exploration of the human proteome is one of the major challenges within life sciences aiming for a better understanding of disease-related processes [1]. Alternatives to widely used mass spectrometric analysis emerge such as developments of miniaturized and parallelized technologies based on affinity interactions and among these methods, various protein microarrays have been implemented into proteomic profiling approaches demonstrating their applicability in high-throughput screening for marker proteins in patient samples [2]. Two protein microarray formats have been developed: (1) For reverse-phase microarrays large numbers of tissue and cell lysates or serum samples are spotted onto functionalized array surfaces for a parallel analysis of a single parameter and (2) a forward-phase setting, which

Helena Bäckvall and Janne Lehtiö (eds.), *The Low Molecular Weight Proteome: Methods and Protocols*,
Methods in Molecular Biology, vol. 1023, DOI 10.1007/978-1-4614-7209-4_8, © Springer Science+Business Media New York 2013

both utilize immobilized capture reagents to analyze many parameters in each of the sample that is profiled in multiplexed sandwich immunoassays or antibody arrays [3].

While dedicated robotic devices are needed to produce planar protein microarrays to arrange molecules on microscope glass slides with functionalized surfaces, alternative platforms can be employed for a parallelized and miniaturized analysis. One of these is based on a flow cytometric readout system to currently differentiate up to 500 color-coded micrometer-sized beads [4]. Arrays are created by suspending beads of different codes in one tube and each bead identity represents an immobilized capture reagent. The platform has recently been utilized to adapt the concept of antibody arrays from previously described planar arrays [5, 6] and multiplexed sandwich immunoassays. The described workflow, summarized in Fig. 1, offers a microtiter plate-based alternative to methods based on planar microarrays for the analysis of labeled serum and plasma protein profiling [7–10] and can be used for highly multiplexing in both the dimension of parameters measured per sample as well as samples studied per analysis. An example of a protein profile obtained from this approach is given in Fig. 2. Here, intensity levels over more than two orders of magnitude and a low-intensity variability of ≤20 % are observed.

2 Materials

2.1 Bead Coupling

1. Beads: MagPlex or MicroPlex microspheres (Luminex Corp).
2. Activation buffer (1×): 100 mM Monobasic Sodium Phosphate (Sigma), pH 6.2, store at +4 °C for up to 3 months and at −20 °C for long term.
3. EDC solution: Prepare aliquots of 1-*ethyl*-3-(3-dimethylamino-propyl) carbodiimide hydrochloride (EDC, Pierce, or Proteochem) in screw-capped tubes and store at +4 °C. Dissolve in activation buffer to 50 mg/ml directly prior usage.
4. S-NHS solution: Prepare aliquots Sulfo-*N*-Hydroxysuccinimide (NHS, Pierce) aliquots in screw-capped tubes and store at +4 °C. Dissolve in activation buffer to 50 mg/ml directly prior usage.
5. Coupling buffer: 100 mM 2-(*N*-morpholino)ethanesulfonic acid (MES) pH 5.0, store at +4 °C for up to 3 months and at −20 °C for long term.
6. Wash buffer: 0.05 % (v/v) Tween20 in 1× PBS pH 7.4 (PBST).
7. Antibody detection solution: *R*-Phycoerythrin modified anti-species antibodies (e.g., Jackson) diluted to 0.25 μg/ml in PBST (*see* **Note 1**).

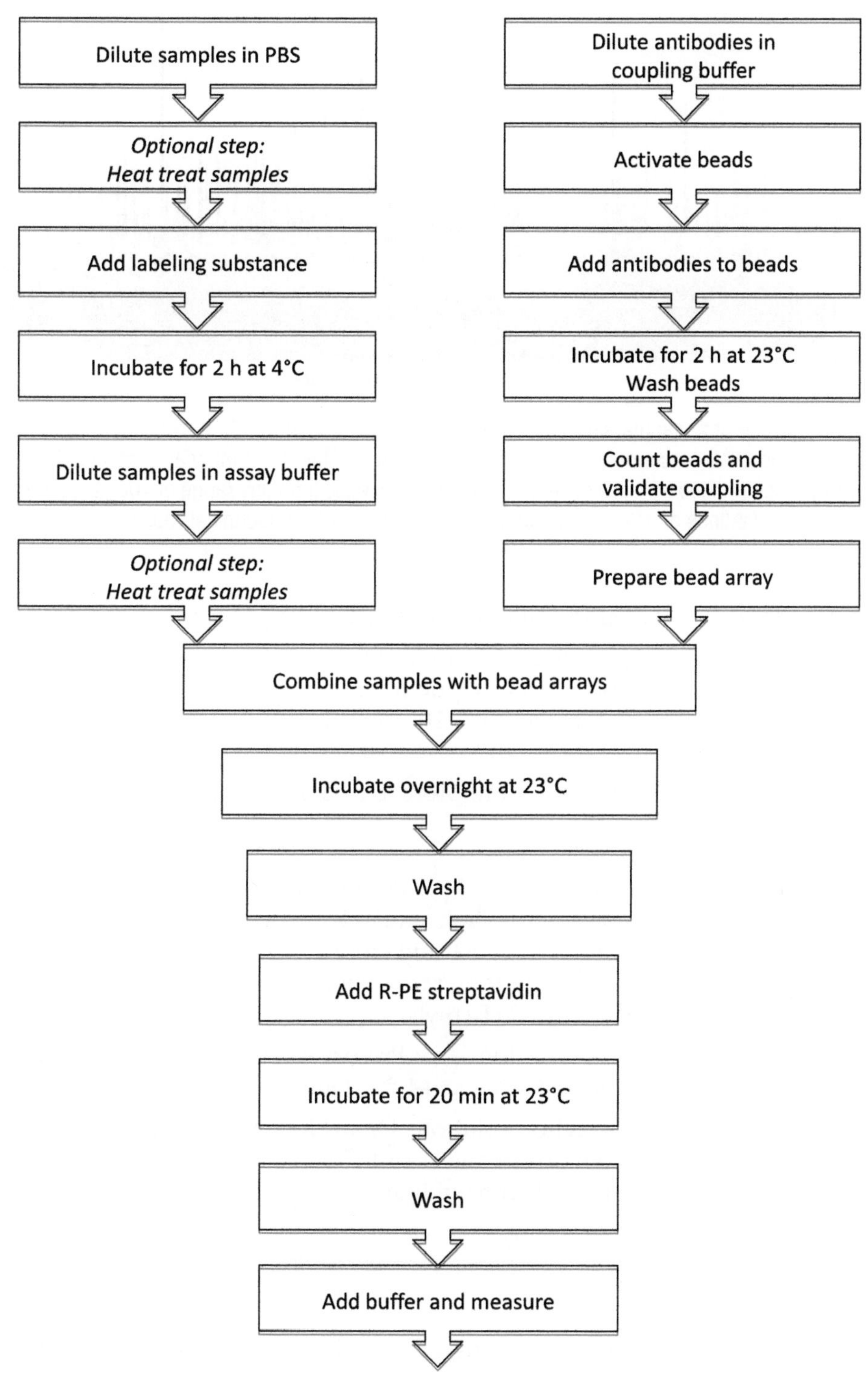

Fig. 1 Experimental workflow

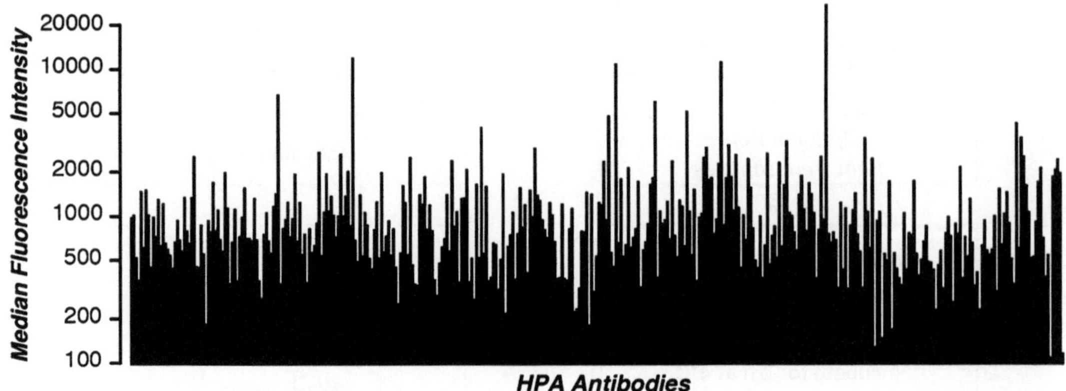

Fig. 2 Intensity profile of a plasma sample. A bead mixture composed of 384 polyclonal antibodies form the Human Protein Atlas (HPA) was employed to generate protein profiles in a biotinylated and heat-treated EDTA plasma sample. Such profiles typically cover a median fluorescence intensity range of 100–20,000 units, while average standard deviations of ≤20 % can commonly be obtained from technical replicates

2.2 Sample Labeling

1. Sample dilution buffer: 1× PBS pH 7.4.
2. Labeling solution: Dissolve Sulfo-*N*-Hydroxysuccinimide-polyethylene oxide biotin (NHS-PEO_4-Biotin, Pierce) directly before use to 10 mg/ml in dimethyl sulfoxide (DMSO, Sigma).
3. Stop solution: Store 1 M Tris–HCl pH 8.0 at +4 °C and add cold.

2.3 Assay Procedure

1. Assay buffer (1×): Prepare 0.1 % (w/v) casein, 0.5 % (w/v) polyvinylalcohol, and 0.8 % (w/v) polyvinylpyrrolidone (all Sigma) in PBST and store at +4 °C for up to 3 months and at −20 °C for long term. Supplement before use with 0.5 mg/ml rabbit IgG (Bethyl).
2. Stop solution (4×): Prepare 4 % paraformaldehyde (PFA) solution and store at +4 °C. Dilute 1:4 in PBS prior to usage.
3. Detection solution: Dilute *R*-Phycoerythrin modified streptavidin (Invitrogen) to 0.5 μg/ml in PBST directly before use and protect from light.

3 Methods

3.1 Bead Coupling

A method for antibody coupling is described for which magnetic and nonmagnetic beads can be utilized. The difference in processing these two bead types refers to the handling of the beads during an exchange of surrounding solutions. When the number of bead identities does not exceed the number of positions found in benchtop microcentrifuges, we suggest to use microcentrifuge tubes or tubes with filter inserts to pellet the beads via centrifugation.

For magnetic beads we suggest to use magnetic forces without centrifugation. For more than 24 couplings in parallel, it is preferred to perform the coupling procedure in a microtiter plate, where proteins can be immobilized on nonmagnetic beads using filter-bottomed microtiter plates (filter pore sizes have to be below bead diameter) and vacuum devices to accommodate plates for to remove liquid via suction. For magnetic bead coupling in plates, dedicated plate magnets are available for up to 384-well plates (LifeSep, Dexter Magnetic Technologies) to facilitate bead sedimentation. It is our experience to preferably use magnetic beads as they facilitate a straightforward integration of the coupling and assay procedure into liquid handling workstations.

1. Prepare antibodies at the desired concentration (e.g., 3 μg or a solution with antibody concentration of 30 μg/ml per 1×10^6 beads) in coupling buffer (*see* **Note 2**).
2. The beads are to be distributed in desired portions (e.g., 80 μl = 1×10^6 beads) into the wells of a half-area plate and the beads are washed with 3× 100 μl activation buffer (*see* **Note 3**).
3. Prepare fresh solutions of NHS and EDC, both at 50 mg/ml in activation buffer. Calculate a use of 0.5 mg of each chemical per bead ID and coupling, and for each bead ID prepare a solution that contains 10 μl NHS, 10 μl EDC, and 80 μl activation buffer.
4. Incubate 20 min under permanent, gentle shaking, and wash thereafter with 3× 100 μl coupling buffer.
5. Continue without interruption (*see* **Note 4**) by adding the antibody solution to the activated beads and incubate for 2 h at +23 °C under permanent, gentle shaking.
6. The beads are washed 3× with 100 μl wash buffer.
7. The beads are then recovered from the wells into microcentrifuge tubes with 3× 100 μl wash buffer. The liquid is removed and 100 μl storage buffer are added prior to the bead storage at +4 °C in the dark for at least 1 h.

3.1.1 Bead Mixture Preparation

The yield of antibody immobilization should be judged after bead coupling. To allow an economic bead consumption and sufficient bead count during the measurements, equal numbers of beads should be combined in a bead mixture. To facilitate this, the beads can be counted and an initial bead concentration can be determined which allows calculating the required volumes to be added in a common stock solution. During this bead counting procedure, the rate of antibody immobilization can be additionally approximated via fluorescently labeled anti-species specific antibodies.

1. The tubes with antibody-coupled beads are to be vortexed and sonicated for 5 min.

2. Each bead solution is diluted 1/100 in antibody detection solution (*see* **Notes 1** and **5**) to a microtiter plate.
3. The plates are incubated for 20 min and measured.
4. The number of counts per bead ID is multiplied by a correction factor of 3.3 for a 1/100 dilution to obtain a first estimation of beads per microliter storage solution. From this number the volumes of beads in storage solution can be calculated which are to be applied into the bead mixture. The required number of beads supplied should be adjusted for each assay procedure and be based on the quantity of beads being counted by the instruments. We suggest to always obtain ≥35 counts per bead ID.
5. After each measurement and for the preparation of new bead mixtures, the count average is to be calculated for each bead ID and new volumes can be determined. We suggest adjusting these volumes to a theoretical bead count, which is 20 % above the estimate: For 100 beads to be counted from the new bead mixture, the previously obtained volumes should be calculated for 120 beads per assay and bead ID.

3.2 Sample Labeling

1. The serum or plasma samples are to be thawed according to the preferred protocol (*see* **Note 6**).
2. The samples are vortexed and centrifuged for 10 min at 10,000 × *g* to pellet insoluble components.
3. A previously designed plate layout, in which samples should be located in random positions, is followed to transfer 30 μl of serum/plasma into the respective wells of a PCR plate, which is then sealed and centrifuged for 2 min at 1,500 × *g*.
4. Transfer 3 μl into a second PCR plate containing 27 μl PBS, seal the plate, vortex, and centrifuge for 2 min at 1,500 × *g*.
5. As an option, the diluted samples are incubated for 30 min at elevated temperatures such as 56 °C (*see* **Note 7**) followed by 15 min at 20 °C, both conducted with a thermo cycler. Using the heated lid function of the cycler helps to prevent the sample liquid to evaporate into the lid/seal.
6. Add 2.5 μl of NHS-Biotin to each well (*see* **Note 8**), and then the plate is sealed, vortexed, centrifuged for 2 min at 1,500 × *g*, and incubated for 2 h at 4 °C under permanent shaking on a microtiter plate mixer.
7. Add 25 μl of 1 M Tris–HCl pH 8.0 to each well, seal the plate, vortex, and centrifuge for 2 min at 1,500 × *g*.
8. Store the plates at −20 °C until usage or use directly.

3.3 Assay Procedure

1. The labeled samples are thawed and diluted 1/50 in assay buffer, which had been prepared in a PCR plate. Seal the plate, vortex, and centrifuge for 2 min at 1,500 × *g*.
2. The samples are incubated for 30 min. As an option, the samples are treated for 30 min at elevated temperatures such as 56 °C (*see* **Note 7**), followed by 10 min at 23 °C using the heated lid function of the thermo cycler. Thereafter, the plate is vortexed and centrifuged for 2 min at 1,500 × *g*.
3. From a previously prepared bead mixture 5 μl are distributed into the wells of 96- or 384-well plates and protected from light. Then 45 μl of the diluted, labeled samples are added to the wells (*see* **Note 5**) and incubated at 23 °C overnight under permanent shaking on a microtiter plate mixer.
4. For 96-well plates, 3× 100 μl wash buffer are used, followed by the incubation with stop solution for 10 min. Thereafter the beads are washed 3× with 100 μl wash buffer. For 384-well plates 60 μl are used.
5. *R*-Phycoerythrin-labeled streptavidin is then added to each well at 0.5 μg/ml and 50 μl and the plates are incubated for 20 min under permanent shaking.
6. The plates are then finally washed 3× with 100 μl wash buffer and 100 μl of wash buffer are added before the plates are measured with the Luminex instrumentation. For 384-well plates 60 μl are used.
7. Set the instrumentation setting according to the bead IDs included in the mixture and count at least 50 beads per bead ID. We suggest to use the "median fluorescence intensity" to further process your data. An example of a plasma protein profile is shown in Fig. 2.

4 Notes

1. Other fluorescent dyes than *R*-Phycoerythrin such as Alexa546, Alexa532, or Cy3 can be utilized as well but Luminex Corp. has indicated that lower reporter signal intensities are to be observed. Different suppliers for *R*-Phycoerythrin can also be compared to achieve a desired assay performance.
2. Employ solutions of purified proteins and avoid other stabilizing proteins, Tris, or other amine-based buffers as they reduce the coupling efficiency.
3. At all times, try to minimize the light exposure, especially to direct sunlight, as the internal fluorescence of the beads as well as reporter fluorophores could be bleached. During incubation, protect the plates with an opaque cover or place plate into a light-tight box.

4. Do not interrupt the activation process after dissolving EDC and NHS, as these active substances are susceptible to hydrolysis resulting in a loss in activity.
5. When combining beads with solutions for counting and assay procedure, always distribute small-volume bead solution (e.g., 5 μl) into the well first and then add larger volume buffer portion (e.g., 45 μl) to allow an instant distribution of the beads.
6. We have found that thawing overnight at +4 °C was most practical if a larger number of samples were to be processed. Otherwise, place tube(s) into a 42 °C water bath until a minor fraction of ice was still visible.
7. We have observed that heat treatment of labeled samples in combination with the applied multiplexed assay procedure allowed to modulate antibody performance [5]. This can lead to improved protein detectability by changing the accessibility of the epitopes in the complex sample solution but should be tested and balanced with the tendency of proteins to precipitate at higher temperatures.
8. Do not interrupt process after dissolving NHS-Biotin, as this active substance is susceptible to hydrolysis resulting in a loss in activity. Add NHS-Biotin to the side of each well using single- or multichannel dispensers so that the labeling reactions for all samples are started contemporaneously.

Acknowledgements

We like to thank the entire staff of the Human Protein Atlas (HPA) for their tremendous efforts within the Human Protein Atlas project. This work is funded by the PRONOVA project (VINNOVA, Swedish Governmental Agency for Innovation Systems), and by grants from the Knut and Alice Wallenberg Foundation and Science for Life Laboratory, Stockholm.

References

1. Hanash S (2003) Disease proteomics. Nature 422:226–232
2. Ayoglu B, Haggmark A, Neiman M, Igel U, Uhlen M, Schwenk JM, Nilsson P (2011) Systematic antibody and antigen-based proteomic profiling with microarrays. Expert Rev Mol Diagn 11:219–234
3. Kingsmore SF (2006) Multiplexed protein measurement: technologies and applications of protein and antibody arrays. Nat Rev Drug Discov 5:310–320
4. Fulton RJ, McDade RL, Smith PL, Kienker LJ, Kettman JR Jr (1997) Advanced multiplexed analysis with the FlowMetrix system. Clin Chem 43:1749–1756
5. Schwenk JM, Gry M, Rimini R, Uhlen M, Nilsson P (2008) Antibody suspension bead arrays within serum proteomics. J Proteome Res 7:3168–3179
6. Rimini R, Schwenk JM, Sundberg M, Sjoberg R, Klevebring D, Gry M, Uhlen M, Nilsson P (2009) Validation of serum protein profiles by a dual antibody array approach. J Proteomics 73:252–266
7. Schwenk JM, Igel U, Kato BS, Nicholson G, Karpe F, Uhlen M, Nilsson P (2010)

Comparative protein profiling of serum and plasma using an antibody suspension bead array approach. Proteomics 10(3):532–540

8. Neiman M, Hedberg JJ, Donnes PR, Schuppe-Koistinen I, Hanschke S, Schindler R, Uhlen M, Schwenk JM, Nilsson P (2011) Plasma profiling reveals human fibulin-1 as candidate marker for renal impairment. J Proteome Res 10:4925–4934
9. Schwenk JM, Igel U, Kato BS, Nicholson G, Karpe F, Uhlen M, Nilsson P (2010) Comparative protein profiling of serum and plasma using an antibody suspension bead array approach. Proteomics 10:532–540
10. Schwenk JM, Igel U, Neiman M, Langen H, Becker C, Bjartell A, Ponten F, Wiklund F, Gronberg H, Nilsson P, Uhlen M (2010) Toward next generation plasma profiling via heat-induced epitope retrieval and array-based assays. Mol Cell Proteomics 9:2497–2507

Part II

Bioinformatics of Proteomics Data

Chapter 9

Protein Quantification by Peptide Quality Control (PQPQ) of Shotgun Proteomics Data

Jenny Forshed

Abstract

This chapter describes how to improve quantitative accuracy and precision in shotgun proteomics by PQPQ (protein quantification by peptide quality control). The method is based on the assumption that the quantitative pattern of peptides derived from one protein will correlate over several samples. Dissonant patterns are assumed to arise either from mismatched peptides or due to the presence of different protein species. PQPQ identifies and excludes outliers and detects the existence of different protein species by correlation analysis. Alternative protein species can then be quantified separately. PQPQ can handle shotgun proteomics data from several MS instruments, data from different kinds of labeling, and label-free data.

We have previously shown that data processing by PQPQ improves the information output from shotgun proteomics by validating the algorithm on seven datasets related to different cancer studies (Forshed et al., Mol Cell Proteomics 10(10):M111.010264, 2011). Data from two labeling procedures and three different instrumental platforms was included in the evaluation. With this unique method using both peptide sequence data and quantitative data, we can improve the quantitative accuracy and precision on the protein level and detect different protein species (Forshed et al., Mol Cell Proteomics 10(10):M111.010264, 2011).

Key words Protein quantification, Data curation, Quantitative accuracy, Quantitative precision, Quantitative shotgun proteomics

1 Introduction

1.1 Background

Very small changes in protein levels can have major effect on the underlying biology why it is very important to be able to detect quantitative differences. Currently, many analyses fail at the point of biological interpretation because of a large quantitative variance and inaccuracy. One part of this problem is that current proteomics data analysis methods are unable to resolve different protein species, such as splice variants or modified subsets of proteins. Since different protein species have different biological functions, it is however essential to be able to detect and quantify those species separately. Today's data analysis output often reports a mean value

Helena Bäckvall and Janne Lehtiö (eds.), *The Low Molecular Weight Proteome: Methods and Protocols*,
Methods in Molecular Biology, vol. 1023, DOI 10.1007/978-1-4614-7209-4_9, © Springer Science+Business Media New York 2013

of different species [1]. The here presented method is unique by using the combination of peptide sequence data and quantitative data to improve the quantitative accuracy and precision of the proteins and detect different protein species in proteomics data.

1.2 Shotgun Proteomics

The identification of the protein components of a biological sample is a complex multistep procedure. In shotgun proteomics, protein samples are digested to peptides by enzymatic cleavage, then typically separated and analyzed in a liquid chromatography-mass spectrometry (LC-MS) system. From the full-scan MS spectrum, precursor peptide ions are selected and fragmented for tandem MS analysis (MS/MS). The fragment ion spectra are interpreted to peptide sequences via a database search and the proteins are inferred from the identified peptides. The protein data output hence rely on several assumptions: perfect tryptic cleavage, that a protein can be identified by only a few peptides, that the peptide-matching algorithms work perfect, and that the protein databases are populated with all proteins and their variants [2]. This is however not true for all proteins in a complex sample. A set of peptides may also be shared by multiple proteins: the protein inference problem. These factors will put uncertainty into the identification and quantification [3]. In a typical shotgun proteomics experiment, a substantial problem is also that low-intensity signals dominate the dataset. Furthermore, many protein identifications are based on only a few peptides, limiting the statistical power in the quantitative results [4]. The presented method called PQPQ is a tool to address some of these issues by curating the peptide data from a shotgun proteomics experiment.

1.3 PQPQ

Obtaining an accurate and precise estimate of the protein ratio from peptide intensities can be done in various ways and no standard methodology is yet defined [4–7]. Several open-source/ academic and commercial software for quantitative analysis of proteomics MS/MS data are available supporting different MS instruments and labeling methods [6, 8–11]. PQPQ is not replacing these methods. PQPQ is a method for quality control and curation of already quantified peptide data. PQPQ then calculates the relative protein quantification based on the cured peptide data. The aim is to increase the quantitative accuracy and precision in the protein output from shotgun proteomics data.

PQPQ compares the quantitative data of every peptide associated with one protein at a time, cross multiple samples, to find outliers and to detect different protein species. Non-correlating peptides are detected as outliers and are excluded. Further, a cluster analysis among the peptides suggests if there exist several species of the protein. If several clusters are found, different protein species are suspected, and are then quantified separately. The method is thoroughly described in the original PQPQ paper [12].

1.4 PQPQ Data Output

PQPQ creates a file with the cured peptide data. This file includes all the information that was originally in the peptide input file plus information of which peptides to include for protein ratio calculation, noted as valid peptides. Also, the results from the cluster analysis are found in this file. The *model peptide* of each cluster is also noted. Further, the output file includes a note if the peptide confidence of the ingoing peptides is too low to support the protein quantification. If the protein has support from less than two peptides, this is also noted. Those notations are also transferred to the protein data output file.

The protein data output file includes the protein ratios: quantitative data and statistics for all proteins with support from assessed peptide data. This file also includes the number of peptides that the quantitative calculations are based on, the standard deviation and the *p*-value for the probability of the ratio to be equal to 1 (student's *t*-test). In case of replicated sample runs, reproducibility (standard deviation between replicates) and number of replicates are reported. If several variants of one protein were detected these are noted. The different variants have unique quantitative outputs.

Further, the PQPQ settings (including normalization factors) are reported and a summary of the number of proteins that were imported, how many were left after filtration, and the number of proteins that were quantified by PQPQ.

1.5 Sample Labeling

PQPQ can handle different types of labeled data, e.g., iTRAQ and TMT, as well as unlabeled data. To find a correlation pattern between peptides, at least four samples (e.g., patients) are required, to get a reliable result. The samples are preferably from the same study, and must be run with equal instrumental setup to be comparable. An ideal experimental setup to meet those conditions is the iTRAQ or TMT labeling. Four, six, or eight samples are then labeled after digestion, mixed, and analyzed together in the mass spectrometer. Technical and laboratory biases after the mixing are consequently "cancelled out." The samples are then compared by quantification of the different reporter ions (each reporter corresponding to a particular sample) in the MS/MS spectra, where the labels diverge in *m/z* position [13, 14].

1.6 System Requirements

- PQPQ can be run on a standard personal computer (*see* **Note 1**).
- MATLAB (The MathWorks Inc., Natick, MA) including the MATLAB statistics toolbox has to be installed on your computer (*see* **Note 2**).
- The PQPQ MATLAB toolbox can be downloaded from www.forshed.se/jenny. The description of PQPQ that is given here refers to PQPQ Version 1. The latest release of the program will always be available at the website or from the author. Each new version of PQPQ also includes an up-to-date manual (*see* **Note 3**).

1.7 Data

PQPQ is designed to handle output data from ProteinPilot™ (Protein Pilot Software 2.0.1.p. (2003–2007), Applied Biosystems/MDS Sciex), Spectrum Mill (Agilent Technologies), and Proteome Discoverer (Thermo Scientific), and can also load manually annotated peptide data as .txt, .csv, .xls, or .xlsx files. The following information is required for PQPQ: the protein accession number(s) associated with the peptides, a value of the peptide confidence, the peptide sequence, the area (or intensity) of the peptide peak (or reporter ion)—one column for each sample—and the corresponding gene name(s) (optional). The input data to PQPQ should include quantitative data from all peptides identified in the samples (not filtered). The data can either be from labeled or label-free experiments. The file shall be structured as one row for each peptide and shall include the columns described in Table 1.

1.8 Input Variables

The following parameters have to be determined before running PQPQ and then be given as input variables to the program:

- *Separate multiple protein IDs*. Having this box ticked, the program will separate proteins identified as a protein group, i.e., proteins that were not possible to separate in the database search of the MS/MS data. The separated identities will then be treated as several entries with the same quantitative information. This can be beneficial if several samples shall be compared. If this box is un-ticked, the joint protein identities will be treated as one.
- *do Normalization (to eq. median)* will normalize each sample column so that each column has the same median peptide intensity (*see* **Note 4**). The normalization factors for each column are found in the protein and peptide output files.
- *The correlation p-value* defines the probability of getting a correlation as large as the observed value by random chance when the true correlation is zero. So, if the *p*-value limit is set to 0.4 (default), the risk of defining a correlation, although there is none, is 40 %. As a guideline, in biological studies with heterogenic samples, a *p-value limit* between 0.4 and 0.1 has shown to work well. How strict you shall be depends on the application.
- *The confident peptide score limit* defines the score limit for which peptides that are defined as highly confident. Different software defines peptide confidence in different ways. Either definition of peptide confidence can be used as long as a high value indicates a high confident. The limit defining a high-scored peptide has to be determined outside the PQPQ software. One way is to define the limit from MAYU [15], where the peptide confidence limit can be determined from the protein false discovery rate (on identification). Another way of defining the score limit is to estimate the FDR of the peptide identification. This can be done by identifying proteins from the MS peptide data in a forward and a reverse database.

Table 1
The data columns required as input to PQPQ

Column information	Vendors' name of the columns		
	Protein Pilot	Spectrum Mill or manually annotated file	Proteome discoverer
The protein accession number(s) associated with the peptide	Accessions	accession_number	Protein accessions
A value of the peptide identification confidence. High number—high confidence	Conf	Score	IonScore *or* XCorr
The peptide sequence	Sequence	Sequence	Sequence
The area/intensity of the peptide peaks, one column for each sample (e.g., iTRAQ label). At least four columns shall be included for good statistics. The columns can be given any name	Area 11X	iTRAQ_1XX	11X
The corresponding gene name(s)	Names	entry_name	
Protein Pilot only for denoting peptides shared between proteins	Annotation		
Proteome Discoverer only for avoiding redundant quantification info. from two detectors (CID and HCD)			Quan Usage

Since the reverse hits are known to be false discoveries, the FDR of the database search can be calculated [16]. The limit can also be set based on recommendation from the data base search.

- *The peptide sum intensity limit* is a limit for how low the sum of the ingoing peptides is allowed to be. If the sum over samples for one peptide is below this limit, the peptide is discarded, and not used in the quantification.
- *Numerator and denominator*. PQPQ calculates ratios (relative quantities) for the protein data output. The numerator and denominator for each ratio are given row by row. Both numerator and denominator can be a mean of several samples.

2 Methods

When installed, a graphical user interface makes PQPQ easy to use on a standard PC. The algorithm is divided into three processes: *Peptide selection*, *Protein ratio calculation*, and *Peptide quantity visualization*. All or one can be chosen, but both Protein ratio calculation and Peptide quantity visualization require that Peptide selection has been done.

2.1 Installation

Save and extract the files in the PQPQ.zip folder into a folder on your computer, typically ...\MATLAB\work\pqpq.

2.2 Example Data

testData2.xlsx is distributed with the algorithm as an example data file. This data is a part of the peptide data detected from a clinical sample by MALDI (Protein Pilot) described in ref. 1.

2.3 Running PQPQ

1. Start MATLAB. Choose "Current Folder" as your PQPQ folder, e.g., C: \MATLAB\work\pqpq.
2. In the Command Window of MATLAB, give the command pqpq.
3. Now the graphical user interface of the program will appear (Fig. 1).

2.4 Inputs

Fill in the required information in the graphical user interface.

- *Path to data folder*. Give the full path to the folder where your peptide data is.
- *Peptide data file name(s)* (.xlsx, .xls, .ssv, .txt). This is the name(s) of the data file in the folder mentioned above. Include the file extension. If the data is in a specific sheet in your excel file, give the excel file and comma the sheet, e.g.,

  peptideData.xls, Sheet1

  In case of sample replicates, put the replicates on single rows, e.g.,

  peptideData.xls, Sheet1

  peptideData.xls, Sheet2
- *Output file name (prefix)* (mandatory). This is the filename that will be given on your result files as prefix, e.g., *filename_peptideData.mat*, *filename_peptideData.xls*, and *filename_proteinData.xls*.
- *Protein names (.txt)* (optional). A data file including the names of the proteins that shall be calculated by PQPQ. It should be in txt format and include the protein accession number names as they are given in the peptide data input file. If no file is given, all proteins in the peptide file are calculated.
- *Choose peptide data format*. *Proteome Discoverer*, *ProteinPilot*, *Spectrum Mill* or *manually annotated* (described above).
- *Do peptide selection*. This is the main function of PQPQ. Having this ticked, the peptide data is cured and all valid peptides for each protein are selected, collected, and saved into the files: *filename_peptideData.mat* and *filename_peptideData.xls*.

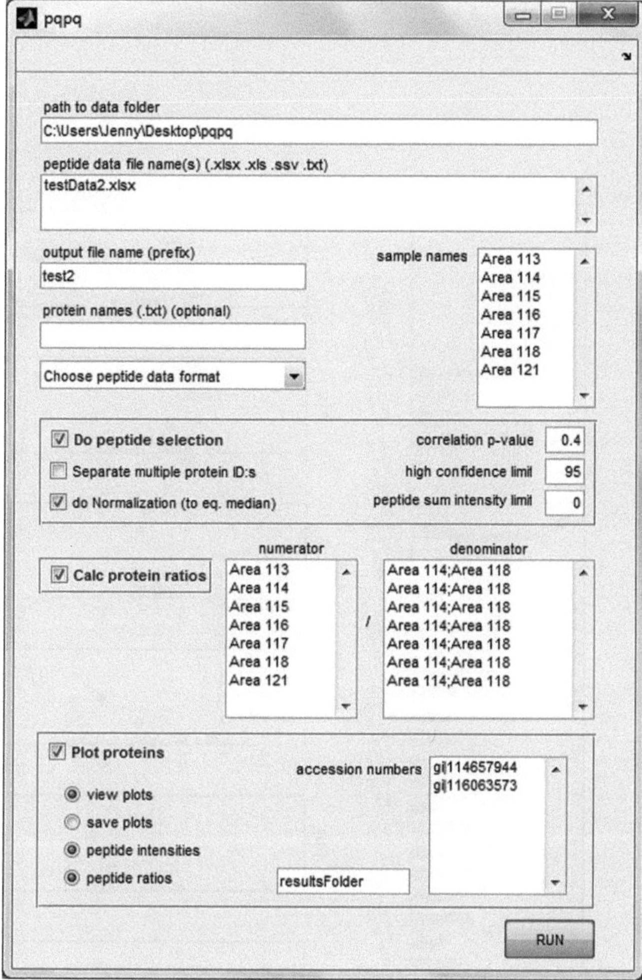

Fig. 1 The graphical user interface of PQPQ

- *Calculate ratios.* Tick if you want to calculate protein ratios. This will give the protein data output file: *filename_proteinData.xls.*
- *Plot proteins.* It can be useful to plot selected proteins after the PQPQ data curation. For example proteins that turn out to come from more than one species can be studied visually by this function. To plot results from a previous run, the following fields are required to be filled in: *path to data folder*, *sample names*, *numerator*, *denominator*, and *output file name.* The function will then load the file *filename_peptideData.mat* and plot the results for each protein given in the list *accession numbers.* The names in the list have to agree with the accession numbers in the peptide data file. Redundant protein names will be plotted only once. The following plot options are possible: *View plots*: The plots will be viewed as MATLAB plots and the user has to press a key to continue and the figure will then

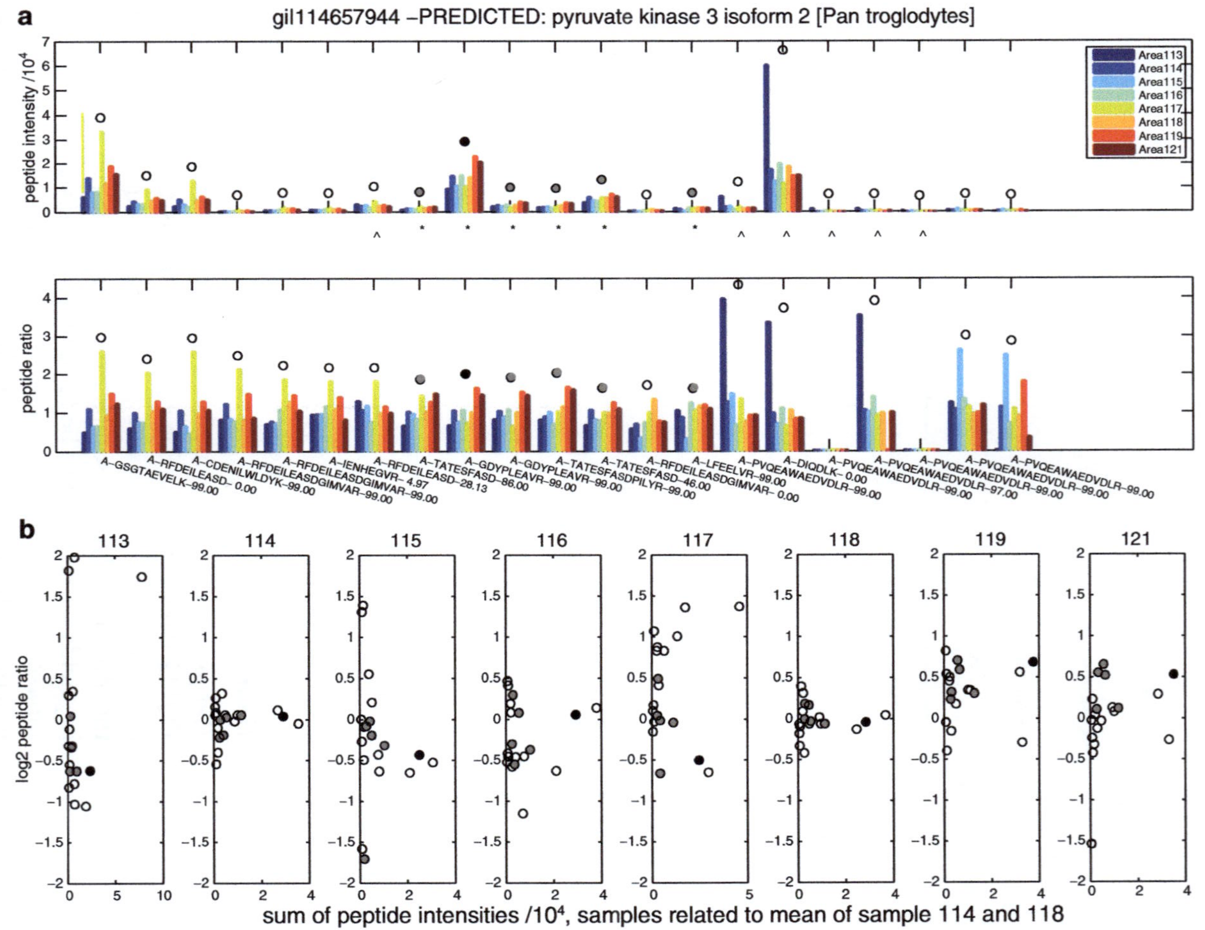

Fig. 2 Figures created with PQPQ. (**a**) The *upper graph* shows the peptide intensities, and *below* is the peptide ratios. Each *color* represents one sample, and each *bar group* is one peptide. The * and ^ denote which peptides that belong to the different variants that PQPQ detected. Two patterns are seen in this case. (**b**) The *bottom graph* shows ratios related to peptide intensity. o = peptide, *grey dots* are included in the quantitative calculations of the main/selected variant, *black dot* marks the selected model peptide in the main variant

automatically be closed. This makes it possible to "scroll" through several figures. Ticking *peptide intensities*, the peptide intensities will be plotted as bars; *peptide ratios*: the peptide ratios will be plotted as bars. An example is shown in Fig. 2.

A folder named *resultsFolder* (or your given name) will be created in the *path to data folder*. For each of the given *accession numbers*, original quantitative data will be reported in an Excel file here. This file includes the peptide sequences, the calculated ratios, and the original intensity values (from the original input data file). If PQPQ has detected several species/variants of the protein, this is given too. This data is useful when investigating the reason why PQPQ has detected several species or outliers. If *save plots* is ticked, the generated plots will be saved in tif format in this folder too.

3 Notes

1. A better processor will reduce the run time. A larger RAM will be required if the input data is large.
2. PQPQ version 1 is developed on MATLAB R2010b.
3. Please cite the paper when using PQPQ:

   Forshed J, Johansson HJ, Pernemalm M, Branca RMM, Sandberg A, Lehtio J. *Enhanced Information Output From Shotgun Proteomics Data by Protein Quantification and Peptide Quality Control (PQPQ)*. Molecular & Cellular Proteomics. 2011 6;10(10)
4. By normalizing so that the medians of the peptide intensities are equal across all samples, we are assuming that the samples included are of similar character and the median of peptide content can thus be expected to be equal.

References

1. Robinson TJ, Dinan MA, Dewhirst M, Garcia-Blanco MA, Pearson JL (2010) SplicerAV: a tool for mining microarray expression data for changes in RNA processing. BMC Bioinformatics 11(1):108
2. Duncan MW, Aebersold R, Caprioli RM (2010) The pros and cons of peptide-centric proteomics. Nat Biotechnol 28(7):659–664
3. Nesvizhskii A, Aebersold R (2005) Interpretation of shotgun proteomic data - The protein inference problem. Mol Cell Proteomics 4(10):1419–1440
4. Karp NA, Huber W, Sadowski PG, Charles PD, Hester SV, Lilley KS (2010) Addressing accuracy and precision issues in iTRAQ quantitation. Mol Cell Proteomics 9(9):1885–1897
5. Boehm AM, Pütz S, Altenhöfer D, Sickmann A, Falk M (2007) Precise protein quantification based on peptide quantification using iTRAQ™. BMC Bioinformatics 8(1):214
6. Lau KW, Jones AR, Swainston N, Siepen JA, Hubbard SJ (2007) Capture and analysis of quantitative proteomic data. Proteomics 7(16):2787–2799
7. Park SK, Venable JD, Xu T, Yates JR (2008) A quantitative analysis software tool for mass spectrometry–based proteomics. Nat Methods 5:319–322

8. Mueller LN, Brusniak M-Y, Mani DR, Aebersold R (2008) An assessment of software solutions for the analysis of mass spectrometry based quantitative proteomics data. J Proteome Res 7(1):51–61
9. Panchaud A, Affolter M, Moreillon P, Kussmann M (2008) Experimental and computational approaches to quantitative proteomics: status quo and outlook. J Proteomics 71(1):19–33
10. D'Ascenzo M, Choe L, Lee KH (2008) iTRAQPak: an R based analysis and visualization package for 8-plex isobaric protein expression data. Brief Funct Genomic Proteomic 7(2):127–135
11. Cox J, Mann M (2008) MaxQuant enables high peptide identification rates, individualized p.p.b.-range mass accuracies and proteome-wide protein quantification. Nat Biotechnol 26(12):1367–1372
12. Forshed J, Johansson HJ, Pernemalm M, Branca RMM, Sandberg A, Lehtio J (2011) Enhanced information output from shotgun proteomics data by protein quantification and peptide quality control (PQPQ). Mol Cell Proteomics 10(10):M111.010264
13. Bantscheff M, Schirle M, Sweetman G, Rick J, Kuster B (2007) Quantitative mass spectrometry in proteomics: a critical review. Anal Bioanal Chem 389:1017–1031
14. Song XM, Bandow J, Sherman J, Baker JD, Brown PW, McDowell MT, Molloy MP (2008) iTRAQ experimental design for plasma biomarker discovery. J Proteome Res 7:2952–2958
15. Reiter L, Claassen M, Schrimpf SP, Jovanovic M, Schmidt A, Buhmann JM et al (2009) Protein identification false discovery rates for very large proteomics data sets generated by tandem mass spectrometry. Mol Cell Proteomics 8(11):2405–2417
16. Elias JE, Gygi SP (2007) Target-decoy search strategy for increased confidence in large-scale protein identifications by mass spectrometry. Nat Methods 4(3):207–214

Part III

Methods Applied in Clinical Research

Chapter 10

Collection and Handling of Blood Specimens for Peptidomics

Harald Tammen and Rudiger Hess

Abstract

Pre-analytical variables can alter the analysis of blood-derived samples. In particular sample collection and specimen preparation can alter the validity of results obtained by modern multiplex assays (e.g., LC-MS). Low-molecular-weight proteins (peptides) as products of proteolytic cleavage events exhibit a close connection to protease activity. Increased or altered activity of proteases during sample collection, specimen generation, sample storage, and processing is mirrored by alterations in abundance of specific peptides. Awareness of clinical practices in medical laboratories and the current knowledge allow for identification of specific variables that affect the results of a peptidomic study. Knowledge of pre-analytical variables is a prerequisite to understand and control their impact.

Key words Blood, Plasma, Serum, Peptidomics, Specimen, Pre-analytical variables

1 Introduction

Peptidomics [1–4] is defined as the systematic, comprehensive, qualitative, and quantitative multiplex (e.g., mass spectrometry) analysis of endogenous peptides in a biological sample at a defined time point and location [5–10]. The term peptide is typically used for polypeptides in the range up to approximately 20 kDa molecular weight. The origin of the term "peptide" (derived from the Greek terms "peptos," meaning digestible, and "poly-," referring to its composition of two or more amino acids) reflects that peptides are usually products of proteolytic processing of larger precursor molecules (e.g., cleavage of activation peptides from zymogens). Proteases [11] can initiate, modulate, and terminate many important biological functions by specific substrate cleavage.

Due to its ease of access and its central role in biological functions, much emphasis has been placed on analyzing blood specimens. Blood can be regarded as a complex liquid tissue comprising cells and extracellular fluid and provides the major link between cells and tissues of an organism. One critically important, and

Helena Bäckvall and Janne Lehtiö (eds.), *The Low Molecular Weight Proteome: Methods and Protocols*, Methods in Molecular Biology, vol. 1023, DOI 10.1007/978-1-4614-7209-4_10, © Springer Science+Business Media New York 2013

often underestimated, factor modulating the likelihood of success in discovering relevant peptides in blood specimens are the pre-analytical procedures used in handling the samples. Pre-analytical procedures comprise the processes prior to the actual analysis of the sample and include steps needed to obtain the primary sample (e.g., blood), and to obtain the analytical specimen (e.g., plasma, serum, cells). Legal or ethical issues (e.g., importance of informed consents) or potential risks of phlebotomy (e.g., bleeding) are not covered in this article.

Pre-analytic procedures can significantly alter the results derived from the analysis of blood-derived samples [12], especially by introducing unwanted systematic biases, which could lead to increased false-positive findings. These procedures comprise the processes prior to the actual analysis of the sample and include steps needed to obtain the primary sample (e.g., blood) and to obtain the analytical specimen (e.g., plasma, serum, cells). Standardization of blood sample collection is a prerequisite to avoid systemic biases [12, 13]. The choice of suitable specimen is also crucial to minimize the occurrence of ex vivo processes (e.g., proteolysis or platelet activation) during specimen collection and preparation [14]. Although serum is one of the most frequently analyzed blood specimens, it bears several caveats for its utility. The generation of serum is associated, e.g., with the activation of coagulation cascade and the complement system. These processes influence the composition of the samples since they result in cell lysis (e.g., thrombocytes, erythrocytes) and protease activation. The activation of proteases leads to the generation of ex vivo-generated peptides. An evident example of peptides as messengers of ex vivo protease activity are the pronounced differences in the peptidome between plasma and serum samples [13, 15, 16]. In serum numerous peptides derived from the coagulation cascade (e.g., activation peptides or other proteolytic events [17]) are present which are not detectable in plasma, since ion-dependent proteases are inhibited by addition of chelate-forming agents. On the other hand a fast degradation of analytes (e.g., hormones) may occur [18]. As a consequence of cell lysis the concentration of components in the extracellular fluid such as aspartate-aminotransferase, serotonin, neuron-specific enolase, and lactate-dehydrogenase is increased [19] in serum. It has also been reported that serum undergoes significant time-dependent changes, which, among other things, affects the reproducibility of sample collection and analysis [20]. A reason to use serum as a specimen is based on the notion that the proteome or the peptidome of serum may reflect biological events [21]. Post-sampling proteolytic cleavage products have been proposed as biomarkers, and it has been further suggested that the serum peptidome is of particular diagnostic value for the detection of cancer [22]. However, it has been reported that more protein changes occur in serum than in plasma [20].

Thus, it can be expected that the reproducibility of such ex vivo proteolytic events is comparatively low.

To counteract the activation of the coagulation cascade citrate, EDTA, or heparin are added to the blood. Citrate and EDTA inhibit coagulation and other enzymatic processes by inhibiting ion-dependent enzymes via chelate formation with ions while heparin inhibits coagulation through the activation of antithrombin III. The main concern associated with heparinized plasma for proteomic studies is that heparin is a polydisperse charged molecule that binds many proteins nonspecifically [23, 24]. Further, heparin may also influence separation procedures and mass spectrometric detection of peptides and small proteins due to its similar molecular weight [16].

In contrast to many of the challenges of using serum as a sample, plasma offers several advantages. Plasma acquisition is less time consuming than the serum acquisition. Separation of cells and the liquid phase, that includes the depletion of thrombocytes below 10 cells/nL to obtain platelet-poor plasma, can be performed subsequent to sample collection since no clotting time (30–60 min) is required. In comparison to serum the amount of plasma generated from blood is approximately 10–20 % higher. Additionally the protein content of plasma is also higher than in serum because of the presence of clotting factors and associated components. Furthermore, proteins may be bound to the clot, resulting in decrease of protein concentration during formation of the clot.

On the peptidomic level more peptides are observed in serum as compared to plasma [13, 15] and this increase results from a differential degree of protease activation in the different specimens. Avoiding ex vivo degradation of proteins and peptides is beneficial, since the complexity of the resulting sample is markedly reduced and for the analysis of in vivo protease cleavage sites, it is mandatory to preserve the integrity of the peptidome.

Conversely it is known that the complement system is also activated in vitro in EDTA and citrate plasma [25]. Furthermore spontaneous clotting and clot formation during storage or during freeze–thaw cycles might occur.

1.1 Collection of Blood Samples

It has been reported that the most frequent faults in the pre-analytical phase are the result of erroneous procedures for sample collection (e.g., blood drawing from an infusive line resulting in sample dilution) [26]. The design of blood collection devices may aid in correct sampling: evacuated containers sustain the draw of the accurate quantity of blood to ensure the correct concentration of additives or the correct dilution of the blood, such as in the case of citrated plasma. The speed of the blood draw is also controlled and restricts the mechanical stress. The favored site of collection is the median cubital vein, which is generally easily found and accessed. As such, it will be most comfortable for the patient and should not evoke additional stress. Preparation of the collection site includes proper cleaning of the skin with an alcohol (2-propanol). The alcohol must be allowed to evaporate, since commingling of remaining alcohol with the blood sample may result in hemolysis, raise the levels of distinct analytes, and cause interferences. The position of the patient (standing, lying, seated) can affect the hematocrit [27] and hence change the concentration of analytes. The tourniquet should be applied 3–4 in. above the site of venipuncture and should be released as soon as blood begins flowing into the collection device. The duration of venous occlusion (>1 min) can affect the sample composition. Prolonged occlusion may result in hemoconcentration, subsequently resulting in an increase in miscellaneous analytes, e.g., total protein levels. Blood should be collected from fasting patients in the morning between 7 and 9 am, because ingestion or circadian rhythms can alter the concentration of analytes considerably (e.g., total protein, hemoglobin, myoglobin).

1.2 Processing of Blood Samples

A quick separation of cells from the plasma is favorable, since cellular constituents may liberate substances that alter the composition of the sample. Generally, it is recommended that plasma and serum are centrifuged with 1,300 to 2,000 × g for 10 min within 30 min, after the collection of the sample. The temperature should generally be 15–24 °C [28], unless recommended differently for distinct analytes like gastrin or A-type natriuretic peptide. Processing at 4 °C appears to be attractive, because enzymatic degradation processes are reduced at low temperatures. However, platelets become activated at low temperatures [29] and release intracellular proteins and enzymes, which affect the sample composition. Thus, processing at low temperatures should be performed only after thrombocytes have been removed. Since one centrifugation step may not be sufficient enough for depletion of platelets below 10 cells/nL a second centrifugation step (2,500 × g for 15 min at room temperature) or a filtration step may be required to obtain platelet-poor plasma. This procedure is only applicable to plasma since platelets in serum are already activated.

1.3 Protease Inhibitors

Protease inhibitors would be attractive, but commonly used protease cocktails may introduce difficulties due to interference with mass spectrometry and formation of covalent bonds with proteins resulting in shifting the isoform pattern [30]. Protease inhibitors have been considered and investigated as additives in proteome research to prevent or slow down proteolytic processes and thereby provide a means of more sensitive detection of markers in blood [31].

Even though protein integrity has been shown to be maintained by addition of 15 commercially available protease inhibitors, the usefulness of protease inhibitors in overall protein stabilization of blood samples remains to be investigated in more detail [12]. The presence of certain protease inhibitors in whole blood is toxic to live cells. Stressed, apoptotic, or necrotic cells release substances, and it may be argued that this impacts the composition of serum or plasma until the cellular and the soluble factions of blood are separated. However careful selection of an appropriate protease inhibitor may solve this problem.

2 Materials

1. 20 gauge needles and the appropriate adapter (e.g., Sarstedt, Nümbrecht, Germany) or a Vacutainer system (BD bioscience, Franklin Lakes, USA).
2. Alcohol (2-propanol) in spray flask.
3. Swabs.
4. Examination gloves.
5. Tourniquet or sphygmomanometer.
6. Blood collection tubes (e.g., Sarstedt, Nümbrecht, Germany).
7. Centrifuge with a swinging bucket rotor (e.g., Sigma 4K15, Sigma Laborzentrifugen, Osterode, Harz).
8. 2 mL cryo-vials.
9. Pipette and tips.

3 Methods

1. Venipuncture a cubital vein using a 20 gauge needle (diameter: 0.9 mm, e.g., butterfly system max. tubing length: 6 cm). If a tourniquet is applied, it should not remain in place for longer than 1 min (risk of falsifying results due to hemoconcentration). As soon as the blood flows into the container, the tourniquet has to be released at least partially. If more time is required, the tourniquet has to be released so that circulation resumes and normal skin color returns to the extremity.
2. Prior to blood collection for proteomics analysis, aspirate blood into a first container (blood collection tube). This is done to flush all surfaces and remove initial traces of contact-induced coagulation. This sample is not useful for analysis.
3. Afterwards, draw blood into a standard EDTA- or citrate-containing syringe. Depending on ease of blood flow, several samples can be collected. Free flow with mild aspiration has to be assured to avoid hemolysis.
4. After venipuncture, obtain plasma by centrifugation for 10 min at 2,000 × *g* at *room temperature*. This centrifugation has to be started within 30 min after blood collection. The resulting plasma sample has now been separated from red and white

blood cells in an efficient and gentle way. Nevertheless, a significant number of platelets (~25 %) are still present in the sample. This requires an additional preparation step.

5. Remove platelet by centrifugation. Transfer the plasma sample into a second vial for another centrifugation for 15 min at 2,500 × *g* at *room temperature*. After this centrifugation, transfer the supernatant in aliquots of 1.5 mL into cryo-vials.
6. Transfer samples to a −80 °C freezer within 30 min. Store at −80 °C. Transport of samples is done on dry ice.

4 Notes

4.1 Frequently Made Mistakes: Blood Withdrawal

1. The patient was not fasting (i.e., had eaten prior to sampling).
2. The blood was drawn from an infusive line.
3. The blood was drawn in another position (e.g., supine, upright).
4. The consumables used were different to those recommended.
5. The expiry date of consumables was already reached.
6. The tubes were not properly filled.
7. The tubes were agitated vigorously (shaken instead of gentle movement to dissolve the anticoagulant).
8. The blood sample tubes were not consistently kept at room temperature.
9. The sample tubes were put on ice or in a refrigerator.

4.2 Frequently Made Mistakes: Lab Handling

1. The centrifugation was delayed more than 30 min after the blood withdrawal.
2. A cooling centrifuge was adjusted below room temperature.
3. The centrifugation speed was wrong (e.g., rounds per minute were set instead of g-force).
4. The centrifugation time was wrong.
5. The removal of blood plasma by pipetting was done without proper caution. Consequently the buffy coat or the red blood cells were churned up.
6. The second centrifugation of recovered plasma samples was delayed after the end of first centrifugation.

4.3 Frequently Made Mistakes: Storage of Samples

1. The storage of samples was delayed.
2. The storage temperatures were above −80 °C.
3. The labelling of sample containers is unreadable or confusable.
4. The attachment of the labels to the sample containers was not sufficient during storage or handling resulting in loss of labels.

4.4 General Recommendations

1. A proper first centrifugation should produce a visible white blood cell layer (buffy coat) between red blood cells and plasma. If not, centrifugation speed or time may be wrong.
2. One should discard plasma that is icteric or exhibit signs of hemolysis. One should check with specialist if this may due to that particular disease.

References

1. Schrader M, Schulz-Knappe P (2001) Peptidomics technologies for human body fluids. Trends Biotechnol 19:55–60
2. Baggerman G, Verleyen P, Clynen E, Huybrechts J, De Loof A, Schoofs L (2004) Peptidomics. J Chromatogr B Analyt Technol Biomed Life Sci 803:3–16
3. Soloviev M, Finch P (2005) Peptidomics, current status. J Chromatogr B Analyt Technol Biomed Life Sci 815:11–24
4. Tammen H, Zucht HD, Budde P (2007) Oncopeptidomics - a commentary on opportunities and limitations. Cancer Lett 249:80–86
5. Schulz-Knappe P, Zucht HD, Heine G, Jurgens M, Hess R, Schrader M (2001) Peptidomics: the comprehensive analysis of peptides in complex biological mixtures. Comb Chem High Throughput Screen 4:207–217
6. Pierson J, Norris JL, Aerni HR, Svenningsson P, Caprioli RM, Andren PE (2004) Molecular profiling of experimental Parkinson's disease: direct analysis of peptides and proteins on brain tissue sections by MALDI mass spectrometry. J Proteome Res 3:289–295
7. Pasinetti GM, Ungar LH, Lange DJ, Yemul S, Deng H, Yuan X et al (2006) Identification of potential CSF biomarkers in ALS. Neurology 66:1218–1222
8. Sasaki K, Sato K, Akiyama Y, Yanagihara K, Oka M, Yamaguchi K (2002) Peptidomics-based approach reveals the secretion of the 29-residue COOH-terminal fragment of the putative tumor suppressor protein DMBT1 from pancreatic adenocarcinoma cell lines. Cancer Res 62:4894–4898
9. Zhang X, Wei D, Yap Y, Li L, Guo S, Chen F (2007) Mass spectrometry-based "omics" technologies in cancer diagnostics. Mass Spectrom Rev 26:403–431
10. Lopez MF, Mikulskis A, Kuzdzal S, Golenko E, Petricoin EF III, Liotta LA et al (2007) A novel, high-throughput workflow for discovery and identification of serum carrier protein-bound Peptide biomarker candidates in ovarian cancer samples. Clin Chem 53:1067–1074
11. Overall CM, Blobel CP (2007) In search of partners: linking extracellular proteases to substrates. Nat Rev Mol Cell Biol 8:245–257
12. Rai AJ, Vitzthum F (2006) Effects of preanalytical variables on peptide and protein measurements in human serum and plasma: implications for clinical proteomics. Expert Rev Proteomics 3:409–426
13. Rai AJ, Gelfand CA, Haywood BC, Warunek DJ, Yi J, Schuchard MD et al (2005) HUPO plasma proteome project specimen collection and handling: towards the standardization of parameters for plasma proteome samples. Proteomics 5:3262–3277
14. Hsieh SY, Chen RK, Pan YH, Lee HL (2006) Systematical evaluation of the effects of sample collection procedures on low-molecular-weight serum/plasma proteome profiling. Proteomics 6:3189–3198
15. Omenn GS, States DJ, Adamski M, Blackwell TW, Menon R, Hermjakob H et al (2005) Overview of the HUPO plasma proteome project: results from the pilot phase with 35 collaborating laboratories and multiple analytical groups, generating a core dataset of 3020 proteins and a publicly-available database. Proteomics 5:3226–3245
16. Tammen H, Schulte I, Hess R, Menzel C, Kellmann M, Mohring T et al (2005) Peptidomic analysis of human blood specimens: comparison between plasma specimens and serum by differential peptide display. Proteomics 5:3414–3422
17. Koomen JM, Li D, Xiao LC, Liu TC, Coombes KR, Abbruzzese J et al (2005) Direct tandem mass spectrometry reveals limitations in protein profiling experiments for plasma biomarker discovery. J Proteome Res 4:972–981
18. Evans MJ, Livesey JH, Ellis MJ, Yandle TG (2001) Effect of anticoagulants and storage temperatures on stability of plasma and serum hormones. Clin Biochem 34:107–112
19. Guder WG, Narayanan S, Wisser H, Zawta B (2003) Samples: from the patient to the laboratory. Wiley-VCH Verlag GmbH & Co. KGaA, Weinheim, Germany

20. Tammen H, Schulte I, Hess R, Menzel C, Kellmann M, Schulz-Knappe P (2005) Prerequisites for peptidomic analysis of blood samples: I. Evaluation of blood specimen qualities and determination of technical performance characteristics. Comb Chem High Throughput Screen 8:725–733
21. Villanueva J, Shaffer DR, Philip J, Chaparro CA, Erdjument-Bromage H, Olshen AB et al (2006) Differential exoprotease activities confer tumor-specific serum peptidome patterns. J Clin Invest 116:271–284
22. Liotta LA, Petricoin EF (2006) Serum peptidome for cancer detection: spinning biologic trash into diagnostic gold. J Clin Invest 116:26–30
23. Landi MT, Caporaso N (1997) Sample collection, processing and storage, IARC Sci Publ 142:223–236
24. Holland NT, Smith MT, Eskenazi B, Bastaki M (2003) Biological sample collection and processing for molecular epidemiological studies. Mutat Res 543:217–234
25. Pfeifer PH, Brems JJ, Brunson M, Hugli TE (2000) Plasma C3a and C4a levels in liver transplant recipients: a longitudinal study. Immunopharmacology 46:163–174
26. Plebani M, Carraro P (1997) Mistakes in a stat laboratory: types and frequency. Clin Chem 43:1348–1351
27. Burtis CA, Ashwood ER (2001) Fundamentals of clinical chemistry. WB Saunders Company, Philadelphia, PA
28. Favaloro EJ, Soltani S, McDonald J (2004) Potential laboratory misdiagnosis of hemophilia and von Willebrand disorder owing to cold activation of blood samples for testing. Am J Clin Pathol 122:686–692
29. Mustard JF, Kinlough-Rathbone RL, Packham MA (1989) Isolation of human platelets from plasma by centrifugation and washing. Methods Enzymol 169:3–11
30. Schuchard MD, Mehigh RJ, Cockrill SL, Lipscomb GT, Stephan JD, Wildsmith J et al (2005) Artifactual isoform profile modification following treatment of human plasma or serum with protease inhibitor, monitored by 2-dimensional electrophoresis and mass spectrometry. Biotechniques 39:239–247
31. Hulmes JD, Bethea D, Ho K, Huang S-P, Ricci DL, Opiteck GJ, Hefta SA (2004) An investigation of plasma collection, stabilization, and storage procedures for proteomic analysis of clinical samples. Humana Press, Totowa, USA

Chapter 11

An Automated RP–SCX Solid-Phase Extraction Procedure for Urinary Peptidomics Biomarker Discovery Studies

Crina I.A. Balog, Rico Derks, Oleg A. Mayboroda, and André M. Deelder

Abstract

Urine represents the most easily obtainable body fluid and consequently one of the most common samples in clinical chemistry. The majority of pathological changes in human organs may well be reflected in urine. In this way, urine analysis can aid in disease diagnosis, treatment monitoring, and prognosis. Currently, the most commonly used method for identification of new urine biomarkers involves centrifugation of the urine sample to collect either the soluble urine proteins or the urinary exosomes followed by 1 or 2 protein purification and separation steps before visualization and finally identification of potential biomarkers, usually by mass spectrometry.

Here we present a generally applicable, rapid, and robust method for screening large number of urine samples, resulting in a broad spectrum of native peptides, as a tool to be used for biomarker discovery. The method combines online sample pretreatment with a well-established mass spectrometric technique. Native peptides are extracted from urine samples on a miniaturized reverse-phase–strong cation exchange cartridge system. As the proper identification of native peptides often requires combination of data acquired on different mass analyzers, we have aimed at a procedure providing us with sufficient material to identify and characterize the differentially expressed markers.

Key words Urine, Peptidomics, RP–SCX purification, Native peptides

1 Introduction

In the last two decades, mass spectrometry-based proteomics has emerged as an indispensable tool of modern biomedical science, but it was not before the publication of Petricoin et al. [1] that clinicians fully became aware of the potential of this new technology. Petricoin and coauthors introduced the concept of protein profiling: the fusion of mass spectrometry with pattern recognition, where specific peak profiles, without the knowledge of individual peak identity, are treated as biomarkers. A number of studies have demonstrated the applicability of serum profiling to a range of medical research

Helena Bäckvall and Janne Lehtiö (eds.), *The Low Molecular Weight Proteome: Methods and Protocols*, Methods in Molecular Biology, vol. 1023, DOI 10.1007/978-1-4614-7209-4_11,

questions, including diagnostics of a variety of cancers [1–6]. Recently, protein profiling has also been adapted for biomarker discovery studies in cerebrospinal fluid (CSF) [7, 8] and urine [9].

In comparison to CSF or even plasma, urine is an easily accessible biofluid and besides the possibility of noninvasive sample collection, it has its special advantages. Most importantly, many peptides and small proteins occur in urine at nearly the same concentrations as in plasma, whereas the total protein concentration is relatively low compared with serum [10]. Thus, the relative enrichment of low-molecular-weight components makes urine an attractive target for a peptide profiling approach. Compared to other biofluids, urine is still less explored. The main reason for this is, probably, that the effects of biological variability are represented stronger in urine. Direct mass spectrometric measurements of peptides in urine are mainly hampered by the high salt content and the presence of high-abundance metabolites such as bilirubin. Methods reported for the extraction of peptides prior to mass spectrometric analysis include solvent extraction [11, 12], solid-phase extraction (SPE) [13, 14], ultrafiltration [15, 16], ultracentrifugation [17, 18], precipitation [18, 19], dialysis [20, 21], or a combination of these techniques [22]. The limitation of these methods is that some of them fail to recover low-molecular-mass peptides (e.g., dialysis, precipitation, ultracentrifugation, and ultrafiltration), while others are not specific for peptides (e.g., solid- and liquid-phase extraction). The idea of solid-phase extraction–strong cation exchange (SPE–SCX) extraction of peptides from urine has been previously described by Cutillas et al. [23] but the limitation of this method is the separate purification steps, hampering the application of this method to large-scale clinical studies.

Here, we present a method for the extraction and subsequent qualitative analysis of urinary peptides that combines rapid automated sample pretreatment with well-established mass spectrometric methods. The method consists of an automatic sample cleanup and fractionation system, using a combination of reverse-phase and strong cation exchange (RP–SCX) cartridges. The use of cartridges instead of columns permits processing of large number of samples. Moreover, the carryover effects, a major problem that can influence the accuracy and precision of high-pressure liquid chromatography (HPLC) and liquid chromatography-mass spectrometry (LC-MS) [24], are substantially reduced in our method by using an SPE system based on exchangeable cartridges followed by the robust and reproducible off-line matrix-assisted laser desorption ionization time-of-flight mass spectrometry (MALDI-ToF MS) measurements.

2 Materials

2.1 Chemicals and Reagents

1. Acetonitrile (CH_3CN), HPLC grade.
2. Potassium phosphate (KH_2PO_4) and potassium chloride (KCl). High-purity buffer salts are readily available from a number of laboratory chemical suppliers.
3. Phosphoric acid and trifluoroacetic acid (TFA).
4. Peptide Calibration Standard I containing a mixture of Angiotensin II, Angiotensin I, Substance P, Bombesin, ACTH clip 1-17, ACTH clip 18-39, Somatostatin 28 and covering a mass range ~1,000–3,200 Da.
5. Bovine serum albumin (BSA) standard digest and 2,5-dihydroxybenzoic acid (DHB).

2.2 Urine Samples

After thawing, centrifuge urine samples for 10 min at 1,500×*g*, 4 °C, for the removal of cellular components. One milliliter of the supernatant was used for subsequent analysis.

2.3 Equipment and Suppliers

1. Prospekt 2 system (Spark Holland).
2. Reverse-phase (RP) cartridge, 10- by 2-mm (Hysphere C_{18} HD-Spark Holland).
3. Strong cation exchange (SCX) cartridge 10- by 2-mm (Isolute-SCX-Spark Holland).
4. MALDI-ToF/ToF instrument, Ultraflex II mass spectrometer (Bruker Daltonics).
5. Ultimate LC system (Dionex), and UV detector. Set the UV absorbance detection at 214 nm.
6. C_{18} PepMap 0.3- by 5-mm trapping column (Dionex).
7. C_{18} PepMap 75-μm by 150-mm column (Dionex).
8. HCT-Ultra iontrap (Bruker Daltonics) equipped with a nano-electrospray ionization source.
9. Millipore C_{18} ZipTips were used for manual desalting of the samples.
10. 96-well C_{18} plate (Isolute-96, C_{18}, 25 mg, Biotage).

2.4 Preparation of Buffers

Buffers should be prepared using high-quality deionized or distilled water. Some buffers may include a proportion of organic solvent (2–30 % acetonitrile) and the solvent should be added after the pH of the buffer has been set (*see* **Note 1**). Buffers are degassed prior to use by 10-min sonication.

2.4.1 Automated Sample Cleanup and Fractionation

1. Buffer 1: HPLC-grade acetonitrile (prewash the RP cartridge).
2. Buffer 2: 2 % acetonitrile adjusted to pH 3 with phosphoric acid (equilibration and loading buffer for the RP cartridge) (*see* **Note 2**).
3. Buffer 3: 10 mM Potassium phosphate adjusted to pH 3 with phosphoric acid (washing buffer for the RP cartridge).
4. Buffer 4: 10 mM Potassium phosphate in 30 % acetonitrile adjusted to pH 3 with phosphoric acid (elution buffer for the RP cartridge and loading buffer for the SCX cartridge).
5. Buffer 5: 10 mM Potassium phosphate in 20 % acetonitrile adjusted to pH 3 with phosphoric acid (washing buffer for the SCX cartridge).
6. Buffer 6: 10 mM Potassium phosphate and 0.1 M potassium chloride in 20 % acetonitrile adjusted to pH 3 with phosphoric acid (elution buffer, low salt concentration, for the SCX cartridge).
7. Buffer 7: 10 mM Potassium phosphate and 0.5 M potassium chloride in 20 % acetonitrile adjusted to pH 3 with phosphoric acid (elution buffer, high salt concentration, for the SCX cartridge).

2.4.2 RP-HPLC/ESI-MS

1. Solvent A: 2 % acetonitrile in 0.1 % formic acid in deionized water.
2. Solvent B: 95 % acetonitrile in 0.1 % formic acid in deionized water.

2.4.3 Sample Desalting for MALDI Measurements

1. Solvent 1: 0.1 % TFA in deionized water (equilibration and washing).
2. Solvent 2: 50 % ACN in 0.1 % TFA (wetting).
3. Solvent 3: 2 g/l DHB in 50 % acetonitrile and 0.1 % TFA (elution).

3 Methods

3.1 Automated Sample Cleanup and Fractionation

1. After thawing, centrifuge the urine samples for 10 min at 1,500 rcf at 4 °C for the removal of cellular components (*see* **Note 3**).
2. Use 1.2 ml of the supernatant for subsequent analysis. Different sample volumes may be used, depending on the research question (*see* **Note 4**).
3. Set up a system (e.g., the Prospekt 2) (*see* **Notes 5** and **6**) to perform automatic sample pre-cleaning and fractionation using two 10 × 2 mm SPE cartridges, a reversed-phase cartridge, and a strong cation exchange cartridge, *see* Fig. 1 [25]. Set the aspiration flow rate at 10 ml/min and the dispensation flow rate at 2 ml/min (*see* **Note 7**).

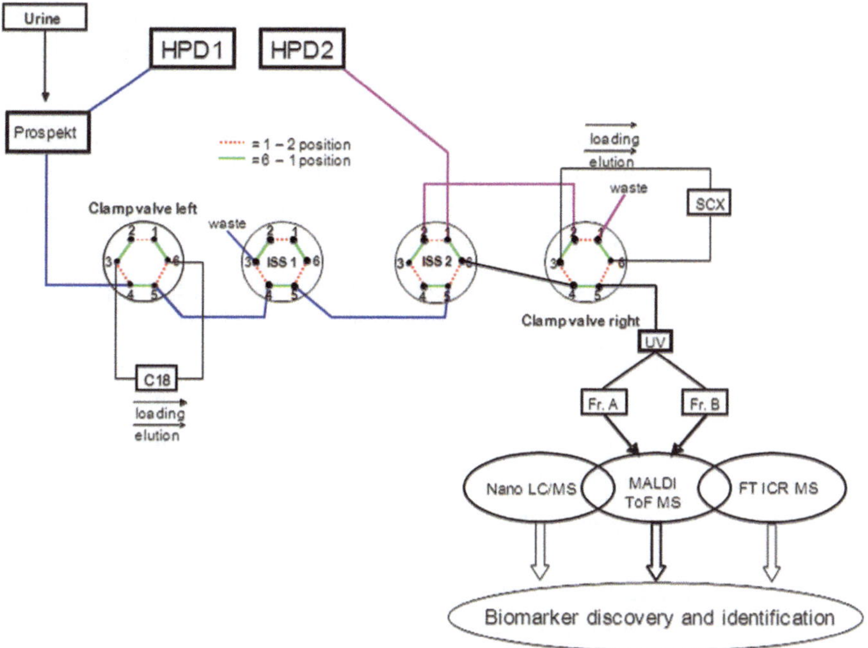

Fig. 1 Workflow for discovery and identification of urinary biomarkers using automatic RP–SCX sample pre-cleaning and fractionation followed by mass spectrometry. Urine samples were loaded on a C_{18} cartridge, desalted, eluted with 30 % acetonitrile, and loaded directly on an SCX cartridge. Peptides were eluted from the SCX cartridge in two fractions (A and B), which were used for initial screening with MALDI-ToF MS. The combination of, e.g., LC-MS/MS fragmentation of the native peptides and high-mass-accuracy data obtained from FT-ICR MS measurements can be used for the subsequent identification of potential biomarkers. Reproduced from ref. 25

4. Wash both cartridges: C_{18} cartridge with 2 ml of buffer 1 and SCX cartridge with 2 ml of buffer 6 (*see* **Note 8**).
5. Equilibrate the C_{18} cartridge with 4 ml of buffer 2 and the SCX cartridge with 2 ml of buffer 5.
6. Apply 1 ml of sample on the C_{18} cartridge (*see* **Notes 4** and **9**).
7. Wash the C_{18} cartridge with 4 ml of buffer 2.
8. Wash the C_{18} cartridge with 2 ml of buffer 3.
9. Elute directly from the C_{18} twice with 1 ml of buffer 4 to the SCX cartridge (*see* **Note 10**).
10. Wash the SCX cartridge twice with 2 ml of buffer 5 (*see* **Note 11**).
11. Elute fraction A with 100 μl of buffer 6 (*see* **Note 12**).
12. Elute fraction B with 100 μl of buffer 7 (*see* **Notes 13** and **14**).
13. Concentrate the two fractions either by freeze-drying overnight or by vacuum centrifugation. The latter approach is preferable, but drying the sample too long should be avoided.
14. Reconstitute the two fractions in 100 μl of 0.1 % TFA. At this stage the samples are ready for further mass spectrometric analysis.

3.2 Qualitative and Quantitative Method Evaluation

The performance of multidimensional sample cleanup and fractionation may be evaluated, e.g., by using a standard digest (BSA) (*see* **Note 15**).

1. Fractionate five BSA digest (8 pmol/ml) replicates on the automatic RP–SCX system.
2. Combine fraction A and B and after drying reconstitute the sample in 100 μl of 0.1 % TFA.
3. Desalt 3 μl of each sample following the ZipTip protocol and analyze the samples using MALDI-ToF MS.
4. Desalt 3 μl of 80 fmol/μl BSA digest following the ZipTip protocol and analyze the sample using MALDI-ToF MS.
5. Compare the sequence coverage of the fractionated and unfractionated BSA digests for the qualitative evaluation (*see* **Note 16**).

Analyze 5 μl of each fraction A and B obtained after RP–SCX fractionation of the five replicate BSA tryptic digests on a nano-LC system (*see* Subheading 3.3) to evaluate the quantitative reproducibility of the procedure.

Following data collection, analyze the UV chromatograms. As with most HPLC analysis software, retention times, peak heights, and peak areas can be calculated. Integrate five randomly selected UV peaks in each LC chromatogram, *see* Fig. 2 [25], and calculate the relative standard deviation (relative SDs) (*see* **Note 17**).

It is important to keep in mind that the quantitative reproducibility of the method should also be tested on a real sample (*see* **Note 18**).

3.3 Nano-LC Sample Analysis

For the nano-flow liquid chromatography preferably use a system with online degassing capability.

1. Inject 5 μl of sample onto a C_{18} PepMap trapping column.
2. Wash with 100 % of solvent A at 20 μl/min for 40 min.
3. Separate the peptides on a C_{18} PepMap column at a constant flow rate of 200 nl/min. The large majority of compounds should elute within the gradient time (start with 10–60 % of solvent B over 50 min).
4. Increase the content of organic solvent up to 90 % of solvent B in 20 min.
5. Wash the column for another 10 min before returning to original conditions.

3.4 Analysis of Fractions Using MALDI-MS

Most forms of modern mass spectrometry can be applied directly to the fractionated material produced using this automated sample cleanup and fractionation. Importantly, since there is sufficient material available after the SPE extraction, the method makes it possible to reanalyze samples with different types of mass spectrometers, thus leading to a higher chance for a positive identification of potential biomarkers. MALDI-ToF MS can be

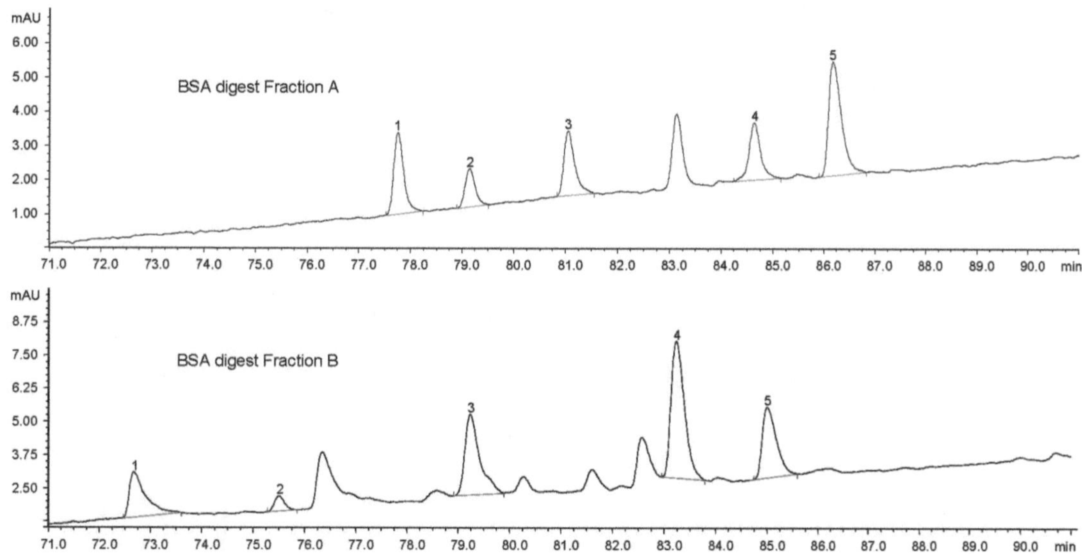

Fig. 2 Nano reversed-phase chromatography of fraction A and B after RP–SCX fractionation of BSA tryptic digests. For the evaluation of the quantitative reproducibility of the automatic RP–SCX pre-cleaning and fractionation system fraction A and B obtained from five BSA tryptic digest replicates were analyzed separately on a nano-LC system using a C_{18} column. A representative chromatogram from both fraction A and B is shown. In each LC chromatogram five randomly selected UV peaks were integrated for fraction A and B. The peak areas were used to calculate the relative standard deviation (relative SD). Reproduced from ref. 25

performed after using the ZipTip protocol for sample desalting, eluting the sample directly on to a MALDI target plate with 1–2 μl of 2 g/l DHB in 50 % of acetonitrile and 0.1 % of TFA (solvent 1). Moreover, for instruments fitted with a reflectron and fragmentation capacity (e.g., LIFT unit for Bruker MALDI-ToF MS instruments) individual peptides may be sequenced and identified, *see* Fig. 3 [25]. The use of an HPLC system coupled to a mass spectrometer with (nano-)electrospray ionization can be applied directly to the eluted fractions (*see* **Note 19**). These types of instrumentation are excellent means to obtain sequence information providing the possibility for identification of the putative biological markers.

When the aim is to analyze a large number of samples the use of "C_{18} well plates," to desalt the fractions prior to MALDI-ToF MS analysis, should be considered.

1. After elution evaporate the organic content using a vacuum centrifuge, dry the eluate, and then reconstitute it in 35 μl of 0.1 % TFA for further analysis.
2. Apply 1 μl of the eluate containing the captured peptides on a stainless steel MALDI target plate and allow to dry.
3. Apply 1 μl alpha-cyano-4-hydroxycinnaminicacid (HCCA) matrix on top of each sample in a concentration of 10 mg/ml in MQ:acetonitrile 1:1 and 0.1 % of TFA (*see* **Note 20**).
4. Collect about 2,000 shots per sample and sum the collected data (*see* **Note 21**).

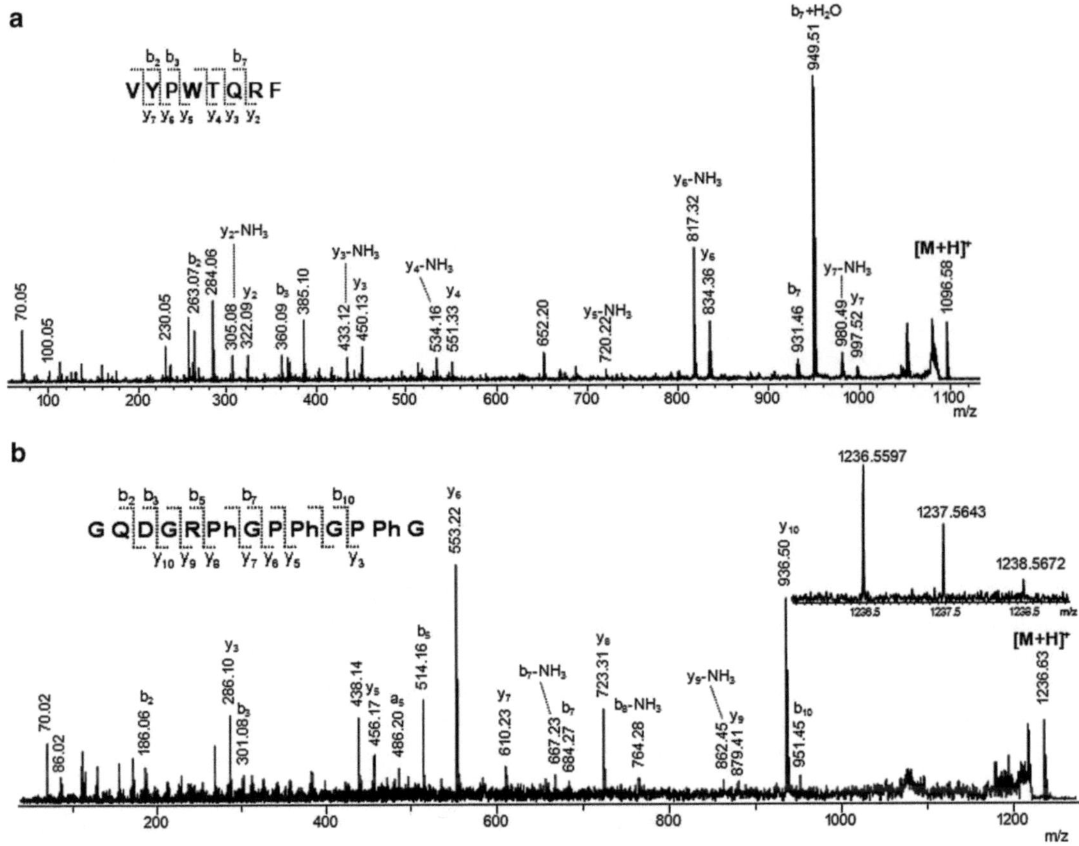

Fig. 3 Identification of potential biomarkers using MALDI-ToF MS/MS and MALDI-FT ICR. (**a**) MALDI-ToF MS/MS of *m/z* 1,096.6 corresponding to the hemoglobin fragment VYPWTQRF; (**b**) MALDI-ToF MS/MS and MALDI-FT ICR MS (*inlet*) analysis of *m/z* 1,236.6 corresponding to the collagen 1A1 fragment GQDGRPhGPPhGPPhG (Ph is hydroxyproline). Reproduced from ref. 25

4 Notes

1. Buffers for cation exchange chromatography should include a proportion of an organic solvent, in our case acetonitrile, to limit the hydrophobic interactions with the exchanger, and the solvent should be added after the pH of the buffer has been set.
2. Elution of peptides from reversed-phase material is generally carried out with an acidic mobile phase. TFA is the most commonly used additive because of its volatility. However, because phosphate buffers were used for the strong cation exchange chromatography, phosphoric acid was used for the pH adjustment.
3. Urine samples can be first void, morning, random catch, or 24 h. First morning urine is usually not recommended in biomarker discovery studies due to its high bacterial content. Usually, to avoid this, a midstream urine is collected which is

more or less free of contamination. However, the selection of urine samples strictly depends on the research question.

4. By replacing the loop of the Prospekt 2 system, other volumes may be injected. The sample vial should always contain 200 μl extra sample for an accurate injection volume (also depending on the shape of the sample vial).

5. The use of off-line cartridges (C_{18} cartridges, e.g., Bakerbond octadecyl SPE cartridges, J.T. Baker; and SCX cartridges, e.g., Sep-Pak AccellPlus SCX, Waters) for manual purification may be a good alternative, if there is no system for the automatic purification. The manual purification should include the same steps as the automatic purification: conditioning of the cartridges, sample application, washing of the cartridges followed by sample elution, as described in the manufacturer's guideline. These steps can be proceeded using vacuum or pressure (nitrogen, air). If a vacuum manifold is available, the cartridges should be placed on the vacuum manifold and the collection tubes in the rack below the output guides. Turn on the vacuum and adjust the pressure to ~7 mmHg using the vacuum control valve. If no control valve is available on your manifold, target a flow rate of about 1–2 drops per second. The same target flow rate should be considered when using pressure. The manual purification is a laborious process, and it requires serious attention. It may take around 30–40 min to process one sample.

6. The automatic purification method is very robust and highly reproducible and it takes about 12 min to process one sample. With this method it is possible to purify about 120 urine samples per day.

7. The Prospekt 2 system has a maximum flow rate of 10 ml/min. As experience has shown, lower flow rates should be used to permit interaction of the peptides with the hydrophobic and cation exchange material. The peptides will thus have the time to bind. The conditioning steps and washing steps may be proceeded with a higher flow rate.

8. All chromatographic and SPE cartridges contain residuals and plasticizers. These should be driven off prior to analyte binding by washing extensively with the elution buffer followed by the equilibration buffer.

9. Unless adjusted, the variable pH, generally observed in urine samples, may lead to variation in recovery during chromatographic fractionation, especially on SCX. Reverse-phase chromatography is less affected by pH variations, and therefore we used this in our method as a first fractionation step. In addition to sample concentration, it combines desalting and pH adjustment required for a subsequent, reproducible fractionation on the SCX cartridge.

10. The smallest peptides and compounds with hydrophilic properties will be easily eluted from the reverse-phase cartridge using 30 % acetonitrile, but the bulk of large proteins will remain on the column. Those large and abundant urinary proteins, which elute at a high concentration of organic modifier, usually obscure the identification of smaller, low-abundance proteins and peptides by mass spectrometry. Moreover, the elimination of large, highly abundant proteins during the RP cleanup step results in an increase in loadability for the native peptides on the SCX cartridge, resulting in improved detection of these peptides during mass spectrometric analysis.
11. The SCX step is necessary because many urinary compounds, such as carboxylic acids and bile salts, have retention times in reversed-phase LC within the range of peptides. Since the p*K*a of the SCX is <1, it is always negatively charged. In acidic solutions, the peptides will be positively charged and thus can be retained by ionic interactions with the SCX bonded phase. This strong ionic retention mechanism allows the sorbent to be extensively washed, which effectively removes anionic (bilirubin) and neutral interferences without seriously affecting the recovery of the peptides.
12. More extensive fractionation increases the number of peptides observed, and depending on the research question it may be decided to use elution buffers with different salt concentrations, increasing the number of collected fractions. Since the method is designed for large studies, requiring a substantial degree of throughput, we decided to use a two-step elution.
13. Since the method has been developed for peptide purification from urine, a maximum of 0.5 M salt concentration of the elution buffer is suitable. If SCX purification is used in combination with C4 reversed-phase material for protein and large peptide purification a higher salt concentration may be needed.
14. The use of high salt concentration could create erosions and plugging. The high-salt mobile phase can precipitate at different parts of the chromatographic system. After use the system should be thoroughly cleaned with water.
15. The use of a standard digest instead of a peptide mixture is recommendable. Peptides with different hydrophobicity, chain length, and charge are needed for a proper method evaluation. Moreover, the use of a protein digest will provide an additional parameter for the method evaluation: sequence coverage.
16. We obtained a very reproducible sequence coverage of mean (SD) 63 % (0 %) ($n=5$) after RP–SCX fractionation of BSA digests compared with 73 % for the unfractionated BSA digest.
17. The calculated relative SDs corresponding to the five peak areas for the two fractions following the protocol described above were 3–7 %, demonstrating good reproducibility of the automatic RP–SCX fractionation of BSA digests.

18. Biological matrices such as urine are known to influence chromatographic steps. Therefore, it is important to evaluate the workflow on the real sample. The integration of the UV peaks within these chromatograms is more challenging due to the increased number of compounds that are co-eluting.
19. The use of reversed-phase purification as the last step before analysis of peptide mixture is necessary for MALDI measurements, but not essential for ESI. If it is decided to analyze the samples by LC/MS only, evaporation of the organic solvent in combination with a very long washing step on the pre-column is sufficient. To avoid contamination of the mass spectrometer and to retain sufficient sensitivity it is recommended to include a 40-min washing step after the peptides were trapped on the pre-column.
20. For MALDI-ToF MS measurements several application techniques of the sample/matrix were tested. We found that subsequent application of sample and matrix was the simplest, fastest, and most reliable method. Currently α-cyano-4-hydroxycinnamic acid (CHCA) is preferred as a MALDI matrix for peptide analysis, because CHCA gives the highest sensitivity and forms a uniform matrix layer on a MALDI plate, which makes it suitable for automated analysis [26]. On the other hand the peptides are apparently more concentrated into the DHB crystals so that the peak intensities increase on MALDI-MS. However, when using DHB, finding good spots for the MALDI-MS analysis remains a drawback and for this reason it is difficult to automate the MALDI-MS measurements with this matrix. We recommend the use of HCCA matrix for automatic measurements and the use of DHB matrix for manual MALDI-MS.
21. Always sum the same amount of laser shots and select as many regions of a spot as possible to ensure high reproducibility for further statistical analysis.

Acknowledgments

This study was supported by the European Union, Sixth framework program, Specific Measures in support of International Cooperation (INCO), contract 517733.

References

1. Petricoin EF, Ardekani AM, Hitt BA, Levine PJ, Fusaro VA, Steinberg SM, Mills GB, Simone C, Fishman DA, Kohn EC, Liotta LA (2002) Use of proteomic patterns in serum to identify ovarian cancer. Lancet 359:572–577
2. Adam BL, Qu Y, Davis JW, Ward MD, Clements MA, Cazares LH, Semmes OJ, Schellhammer PF, Yasui Y, Feng Z, Wright GL Jr (2002) Serum protein fingerprinting coupled with a pattern-matching algorithm distinguishes prostate cancer from benign prostate hyperplasia and healthy men. Cancer Res 62:3609–3614
3. De Noo ME, Deelder A, van der Werff M, Ozalp A, Mertens B, Tollenaar R (2006) MALDI-TOF serum protein profiling for the detection of breast cancer. Onkologie 29:501–506
4. De Noo ME, Mertens BJ, Ozalp A, Bladergroen MR, van der Werff MP, van de Velde CJ,

Deelder AM, Tollenaar RA (2006) Detection of colorectal cancer using MALDI-TOF serum protein profiling. Eur J Cancer 42:1068–1076

5. Rai AJ, Zhang Z, Rosenzweig J, Shih I, Pham T, Fung ET, Sokoll LJ, Chan DW (2002) Proteomic approaches to tumor marker discovery. Arch Pathol Lab Med 126:1518–1526
6. Yanagisawa K, Shyr Y, Xu BJ, Massion PP, Larsen PH, White BC, Roberts JR, Edgerton M, Gonzalez A, Nadaf S, Moore JH, Caprioli RM, Carbone DP (2003) Proteomic patterns of tumour subsets in non-small-cell lung cancer. Lancet 362:433–439
7. Davidsson P, Sjogren M (2005) The use of proteomics in biomarker discovery in neurodegenerative diseases. Dis Markers 21:81–92
8. Gineste C, Ho L, Pompl P, Bianchi M, Pasinetti GM (2003) High-throughput proteomics and protein biomarker discovery in an experimental model of inflammatory hyperalgesia: effects of nimesulide. Drugs 63(Suppl 1):23–29
9. Vlahou A, Schellhammer PF, Mendrinos S, Patel K, Kondylis FI, Gong L, Nasim S, Wright JG Jr (2001) Development of a novel proteomic approach for the detection of transitional cell carcinoma of the bladder in urine. American Journal of Pathoogy 158:1491–1502
10. Hortin GL, Sviridov D (2007) Diagnostic potential for urinary proteomics. Pharmacogenomics 8:237–255
11. Kim MK, Warren TC, Kimball ES (1985) Purification and characterization of a low molecular weight transforming growth factor from the urine of melanoma patients. J Biol Chem 260:9237–9243
12. Meng QC, Balcells E, Dell'Italia L, Durand J, Oparil S (1995) Sensitive method for quantitation of angiotensin-converting enzyme (ACE) activity in tissue. Biochem Pharmacol 50:1445–1450
13. Sato K, Hirata Y, Imai T, Iwashina M, Marumo F (1995) Characterization of immunoreactive adrenomedullin in human plasma and urine. Life Sci 57:189–194
14. Walters AE, Myrdal PB, Pinsuwan S, Manka AM, Yalkowsky SH (1997) Determination of melanotan II in rabbit urine using solid-phase extraction sample preparation followed by reversed-phase high-performance liquid chromatography. J Chromatogr B Biomed Sci Appl 690:99–103
15. Fierens C, Thienpont LM, Stockl D, Willekens E, De Leenheer AP (2000) Quantitative analysis of urinary C-peptide by liquid chromatography-tandem mass spectrometry with a stable isotopically labelled internal standard. J Chromatogr A 896:275–278
16. Sanchez JC, Hochstrasser DF (1999) Preparation and solubilization of body fluids for 2-D. Methods Mol Biol 112:87–93
17. McKee JA, Kumar S, Ecelbarger CA, Fernandez-Llama P, Terris J, Knepper MA (2000) Detection of Na(+) transporter proteins in urine. J Am Soc Nephrol 11:2128–2132
18. Thongboonkerd V, McLeish KR, Arthur JM, Klein JB (2002) Proteomic analysis of normal human urinary proteins isolated by acetone precipitation or ultracentrifugation. Kidney Int 62:1461–1469
19. Marshall T, Williams KM (1998) Clinical analysis of human urinary proteins using high resolution electrophoretic methods. Electrophoresis 19:1752–1770
20. Cutler P, Bell DJ, Birrell HC, Connelly JC, Connor SC, Holmes E, Mitchell BC, Monte SY, Neville BA, Pickford R, Polley S, Schneider K, Skehel JM (1999) An integrated proteomic approach to studying glomerular nephrotoxicity. Electrophoresis 20:3647–3658
21. Ostergaard M, Wolf H, Orntoft TF, Celis JE (1999) Psoriasin (S100A7): a putative urinary marker for the follow-up of patients with bladder squamous cell carcinomas. Electrophoresis 20:349–354
22. Heine G, Raida M, Forssmann WG (1997) Mapping of peptides and protein fragments in human urine using liquid chromatography-mass spectrometry. J Chromatogr A 776:117–124
23. Cutillas PR, Norden AG, Cramer R, Burlingame AL, Unwin RJ (2003) Detection and analysis of urinary peptides by on-line liquid chromatography and mass spectrometry: application to patients with renal Fanconi syndrome. Clinical Science (London) 104:483–490
24. Hughes NC, Wong EY, Fan J, Bajaj N (2007) Determination of carryover and contamination for mass spectrometry-based chromatographic assays. The American Association of Pharmaceutical Scientists journal 9:E353–E360
25. Balog CI, Hensbergen PJ, Derks R, Verweij JJ, van Dam GJ, Vennervald BJ, Deelder AM, Mayboroda OA (2009) Novel automated biomarker discovery work flow for urinary peptidomics. Clin Chem 55:117–125
26. Beavis RC, Chaudhary T, Chait BT (1992) Alpha-cyano-4-hydroxycinnamic acid as a matrix for matrix-assisted laser desorption mass-spectrometry. Organic Mass Spectrometry 27:156–158

Chapter 12

Application of Phage Display for Ligand Peptidomics to Identify Peptide Ligands Binding to AQP2-Expressing Membrane Fractions

Byung-Heon Lee and Tae-Hwan Kwon

Abstract

In vitro phage display represents an emerging and innovative technology for the rapid isolation of high-affinity peptide ligands. Phage display technologies using phages comprising a vast library of peptides have become fundamental to the isolation of high-affinity binding ligands for diagnostic and therapeutic applications, e.g., ligand proteomics, discovery of novel protein–protein interactions, antibody engineering, targeted delivery of therapeutic agents, and development of imaging probes. This chapter describes the procedures for phage display selection of peptide ligands that selectively bind to aquaporin-2-expressing membrane fractions of rat kidney.

Key words Aquaporin, Collecting duct, Ligand peptidomics, In vitro phage display, Vasopressin

1 Introduction

Regulation of the water permeability of the apical plasma membrane in the kidney collecting duct principal cells is critical for regulation of renal water reabsorption and body water balance [1]. Aquaporin-2 (AQP2), the vasopressin-regulated water channel protein, is expressed in the apical plasma membrane and subapical vesicles in the collecting duct principal cells and is the chief target for regulation of the osmotic water permeability in response to vasopressin [2, 3]. Acute regulation involves vasopressin-induced trafficking of AQP2 between an intracellular reservoir in vesicles and the apical plasma membrane [4]. Moreover, AQP2 is involved in chronic control of body water balance, which is achieved through regulation of AQP2 expression [5]. Importantly, multiple studies have now underscored a critical role of AQP2 in several inherited and acquired water balance disorders. This includes inherited forms of nephrogenic diabetes insipidus (NDI), acquired states of NDI, and water retention disorders such as congestive heart failure and liver cirrhosis [3]. Consistent with this, a recent study of collecting duct-specific AQP2

Helena Bäckvall and Janne Lehtiö (eds.), *The Low Molecular Weight Proteome: Methods and Protocols*, Methods in Molecular Biology, vol. 1023, DOI 10.1007/978-1-4614-7209-4_12,

knockout mice further confirmed an essential role of AQP2 in the urinary concentration and body water balance [6].

We previously demonstrated high-affinity peptide ligands binding to the AQP2-expressing plasma membrane (PM) fractions and AQP2-expressing intracellular vesicle (ICV) fractions from rat kidney by exploiting in vitro phage display technique [7]. Seven phage clones of high frequency were selected, which showed high affinity to the AQP2-containing PM and/or ICV fractions compared with the binding of nonrecombinant T7 insertless phage clone. Moreover, these phage clones showed significantly lower affinity to intercalated cell H^+-ATPase (B1-subunit) containing membrane fractions, indicating the selectivity of these phage clones to the AQP2-expressing membrane fractions in the collecting duct principal cells. Fluorescein-conjugated peptide labeling also suggested high affinity to the collecting duct cells with subcellular localization of the intracellular compartment and plasma membrane of primary cultured inner medullary collecting duct (IMCD) cells. And library analysis identified proteins having homologous motifs to each high-affinity peptide ligand. Potential protein candidates are searched based on the homologous motifs to the sequence of isolated high-affinity binding peptide ligands, albeit with a high probability of a random match due to short peptide sequences. These proteins are hypothesized to play a potential role in the vasopressin-induced intracellular trafficking of AQP2 and regulation of AQP2 expression, presumably through functional protein–protein interactions or other signaling mechanisms [7].

2 Materials

2.1 Preparation for Plasma Membrane Fractions and Intracellular Vesicle Fractions from Rat Kidney

1. Dissecting buffer for whole-kidney homogenate: 0.3 M sucrose, 25 mM imidazole, 1 mM EDTA, pH 7.2, protease inhibitors (8.5 μM leupeptin and 1 mM phenylmethylsulfonyl fluoride).
2. Homogenizer: IKA® T10 basic HOMOGENIZER WORK-CENTER 3240000S.
3. Ultracentrifugation: Beckman Optima™ L-100XP with Ti-90 rotor.

2.2 Immunoisolation of AQP2-Expressing Plasma Membrane Fractions and AQP2-Containing Intracellular Vesicles Fractions from Whole Kidney in Rat

1. Magnetic beads (Dynal M-280; Dynal Biotech ASA, Oslo, Norway, pre-coated with anti-rabbit IgG).
2. Laemmli sample buffer: 10 mM Tris, 1.5 % SDS, 6 % glycerol, pH 6.8.
3. Primary antibody: An affinity-purified anti-rat AQP2 antibody.

2.3 Screening of Phage Library for AQP2-Expressing Membrane Fraction-Specific Peptides (Biopanning of a T7 Phage Library)

1. A phage library based on T7 415-1b phage vector displaying CX7C (C, cysteine; X, any amino acid residue) random peptides (*see* **Note 1**). Store at −80 °C.
2. BL21 strain of *E. coli* (Novagen). Store at −80 °C.
3. LB medium: 10 g Bacto-Tryptone, 5 g yeast extract, and 5 g NaCl per liter. Autoclave and store at 4 °C.
4. LB agar plates: LB medium + 15 g agar per liter. Store at 4 °C.
5. Top agarose: 1 g Bacto-Tryptone, 0.5 g yeast extract, 0.5 g NaCl, and 0.6 g agarose per liter. Autoclave and store at 4 °C.
6. M9/LB medium: 10 g Bacto-Tryptone, 5 g yeast extract, 10 g NaCl, 1 g NH_4Cl, 3 g KH_2PO_4, 3 g $NaHPO_4{\cdot}7H_2O$ per liter containing 0.4 % glucose, and 1 mM $MgSO_4$.
7. Fetal bovine serum (FBS).
8. TBS: Tris 3.0285 g, NaCl 4.3838 g/500 ml, pH 7.5.

2.4 PCR, DNA Sequencing, and Peptide Sequence Analysis

1. Primers: Up primer (5′-AGCGGACCAGATTATCGCTA-3′) and down primer (5′-AACCCCTCAAGACCCGTTTA-3′).
2. PCR reagents.
3. 96-well reaction plates with round bottom.
4. 2–4 % agarose gel: A solidified gel made by boiling agarose dissolved in TAE buffer (Tris–acetate–EDTA buffer).

2.5 Primary Culture of Rat Kidney Inner Medullary Collecting Duct Cells

1. Cold D-PBS: 1,000 ml ddH_2O, 0.2 g KCl, 0.2 g KH_2PO_4, 1.15 g Na_2HPO_4, 8 g NaCl, 80 mM urea, 130 mM NaCl, pH 7.4, 640 mOsm/kg H_2O.
2. Enzyme solution: 10 ml Dulbecco's modified Eagle's medium/F12 without phenol red, collagenase B, 7 mg of hyaluronidase, 80 mM urea, 130 mM NaCl.
3. Hypertonic culture medium: Dulbecco's modified Eagle's medium/F12 without phenol red, 80 mM urea, 130 mM NaCl, 10 mM HEPES, 2 mM l-glutamine, penicillin/streptomycin 10,000 units/ml, 50 nM hydrocortisone, 5 pM 3,3,5-triiodo-thyronine, 1 nM sodium selenate, 5 mg/l transferrin, 10 % FBS, pH 7.4, 640 mOsm/kg H_2O.
4. Human fibronectin-coated chamber slides.

2.6 Labeling of Cells with Fluorescein-Conjugated Peptides

1. Fluorescein-labeled peptides (*see* **Note 2**).
2. Control peptide (NSSVDK) conjugated to FITC. Protect from light.
3. Fixative: 2.5 % paraformaldehyde in PBS, pH 7.4.
4. Detergent for cell permeabilization: 0.3 % Triton X-100 in PBS.
5. Hydrophilic mounting media containing antifading reagent.
6. Microscope and charge-coupled device (CCD) camera.

3 Methods

3.1 Preparation for Plasma Membrane Fractions and Intracellular Vesicle Fractions from Rat Kidney

1. Anesthetize pathogen-free male Sprague-Dawley rats (200–250 g) under light enflurane inhalation and rapidly remove both kidneys.
2. Place whole kidneys on an ice-cold Petri dish, dissect, and homogenize in 10 ml of dissecting buffer.
3. Centrifuge the homogenate at 4,000 × *g* for 15 min at 4 °C to remove nuclei, mitochondria, and any remaining large cellular fragments. Collect the supernatants, and prepare consecutively low-speed (LS) plasma membrane fractions and high-speed (HS) intracellular vesicle fractions by differential centrifugation of the supernatant at 17,000 × *g* for 30 min (LS) and 200,000 × *g* for 1 h (HS).
4. LS pellet represents fractions enriched for plasma membrane and HS pellet represents fractions enriched for ICV. Resuspend the pellets in dissecting buffer.

3.2 Immunoisolation of AQP2-Expressing Plasma Membrane Fractions and AQP2-Containing Intracellular Vesicles Fractions from Whole Kidney in Rat

1. Prepare membrane fractions enriched either for PM (LS) or ICV (HS) from rat whole kidney.
2. Incubate both magnetic beads and an affinity-purified anti-AQP2 antibody (H7661AP, ~2 μg/10^7) beads with either the PM fractions or the ICV fractions overnight at 4 °C with continuous agitation.
3. For immunoblotting to detect AQP2, carefully wash samples in 0.1 % BSA in PBS three times for 10 min each and the complex of magnetic beads + AQP2 antibody + PM or ICV fractions is separated magnetically. Mix the pellets (magnetically isolated) with Laemmli sample buffer, and follow by heating to 60 °C for 15 min to solubilize proteins. The beads are then removed magnetically, and immunoisolated samples are used for immunoblotting.
4. For in vitro phage display, carefully wash samples in 0.1 % BSA in PBS three times for 10 min each to remove unbound protein. And the complex (i.e., magnetic beads + AQP2 antibody + PM or ICV fractions) is obtained magnetically.

3.3 Screening of Phage Library for AQP2-Expressing Membrane Fraction-Specific Peptides (Biopanning of a T7 Phage Library)

1. Incubate the T7 phage library (a diversity of ~5 × 10^8 pfu) with magnetic beads which are pre-coated with anti-rabbit IgG antibodies. The phages which bound to magnetic beads nonspecifically are removed magnetically, and then use the remaining unbound phage library for the incubation.
2. Rotate gently the T7 phage library at 4 °C overnight to allow to bind to the complex (i.e., magnetic beads + AQP2 antibody + PM or ICV fractions).

3. Next day, wash ten times with 1 ml of M9/LB medium to remove nonspecifically bound phages. The phages bound to the complex are eluted and will be subjected to a plaque assay for counting the number of phage clone.
4. Moreover, these phages will be used to infect a log-phase culture of *E. coli* for amplification. The amplified phages are again subjected to the binding to newly prepared complex of magnetic beads + AQP2 antibody + PM or ICV fractions by an identical procedure.
5. After three rounds of panning in the same manner (*see* **Note 3**) [7, 8], randomly pick up a number of plaques from the PM fractions and the ICV fractions from LB plates. Store each phage clone in 10 μl TBS and sequence.
6. Choose phage clones that reveal amino acid sequences of high frequency and include similar ones listed alongside. Examine high-affinity binding of these phage clones onto the AQP2-immunoisolated fractions (both PM and ICV fractions), relative to the one by T7 insertless phage clone which does not display peptide ligands on the surface of the phage particle.
7. Moreover, to validate the high affinity of these phage clones to the AQP2-expressing PM and/or ICV fractions, perform a binding assay of selected phage clones to the H^+-ATPase (B1-subunit)-expressing PM and/or ICV of the whole kidney.
8. The procedures for biopanning and validation of the phage library of a T7 phage library are shown in Figs. 1 and 2.

3.4 PCR, DNA Sequencing, and Peptide Sequence Analysis

1. After screening, pick a number of plaques randomly and suspend in 10 μl of Tris-buffered saline (TBS) at each well of 96-well reaction plates. Subject all to DNA sequencing.
2. The insert coding region of selected phage clones is amplified by PCR. Prepare 2 μl of the primer pair solution containing 5 pmol/μl of each primer and add to 22 μl of PCR premix. Add 1 μl of phage suspension to each of the PCR reaction mixture containing primers. Run the PCR reaction using the following condition: 35 cycles of 94 °C for 50 s, 50 °C for 1 min, and 72 °C for 1 min; hold at 72 °C for 6 min; and hold at 4 °C until ready to sequence.
3. Check the PCR products (~250 bp) by electrophoresis on a 2–4 % agarose gel, purify, and sequence by an automatic DNA sequencer.
4. Align the deduced amino acid sequences using CLUSTAL W program to find out the consensus sequence. Conduct the NCBI BLAST search against the SWISSPROT database, using the option for short nearly exact matches, to find proteins with significant homology to a peptide sequence (*see* **Note 4**) [7–9]. Choose candidate peptides and synthesize for further study (*see* **Note 5**).

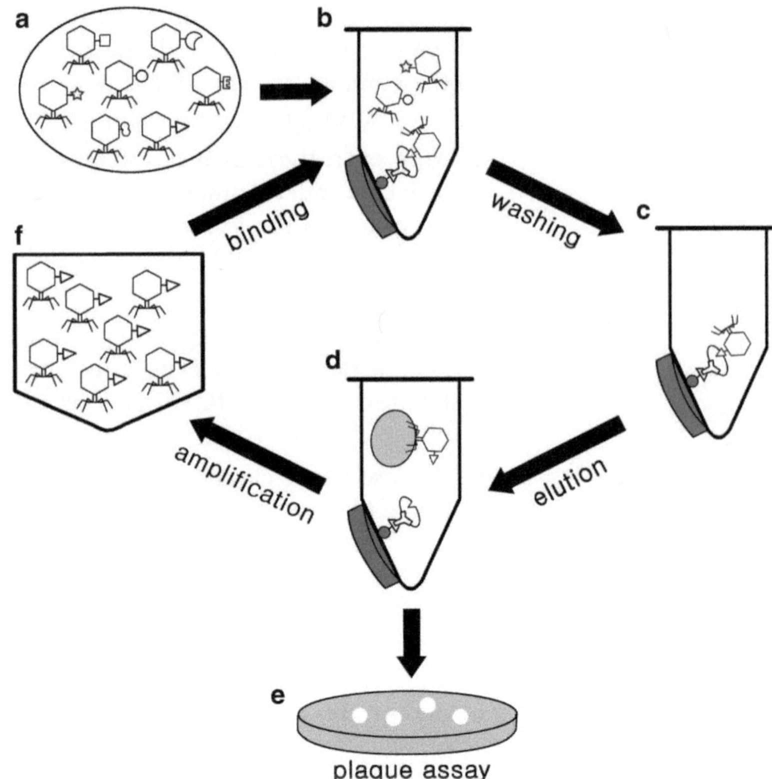

Fig. 1 Flow diagrams for biopanning of T7 phage and plaque assay. (**a**, **b**) An aliquot (1×10^{11} pfu) of the T7 phage library is allowed to bind to the complex (i.e., magnetic beads + AQP2 antibody + PM or ICV fractions). (**c**–**e**) After washing to remove nonspecifically bound phages (**c**), the phages bound to the complex are eluted (**d**) and subjected to a plaque assay for counting the number of phage clone (**e**). (**f**) These phages are used to infect a log-phase culture of *E. coli* for amplification. The amplified phages are again subjected to the binding to newly prepared complex of magnetic beads + AQP2 antibody + PM or ICV fractions by the identical procedure. Three rounds of panning are performed in the same manner [7]

3.5 Primary Culture of Rat Kidney Inner Medullary Collecting Duct Cells

1. Keep male Sprague-Dawley rats (200–270 g) under light enflurane inhalation anesthesia and rapidly remove both kidneys. Place kidneys in cold D-PBS and quickly dissect into inner medulla and other parts [10].
2. Place the inner medulla on an ice-cold Petri dish, mince, transfer to enzyme solution, and incubate at 37 °C under continuous agitation (300 rpm) for 90 min in a humidified incubator (5 % CO_2 and 95 % O_2).
3. Centrifuge the resulting suspension at 160 g for 1 min to get pellet which contained IMCD fragment and IMCD cells.
4. Wash the pellet in pre-warmed culture medium without enzyme and seed the IMCD cell suspension in human fibronectin-coated chamber slides for labeling of fluorescein-conjugated peptides.

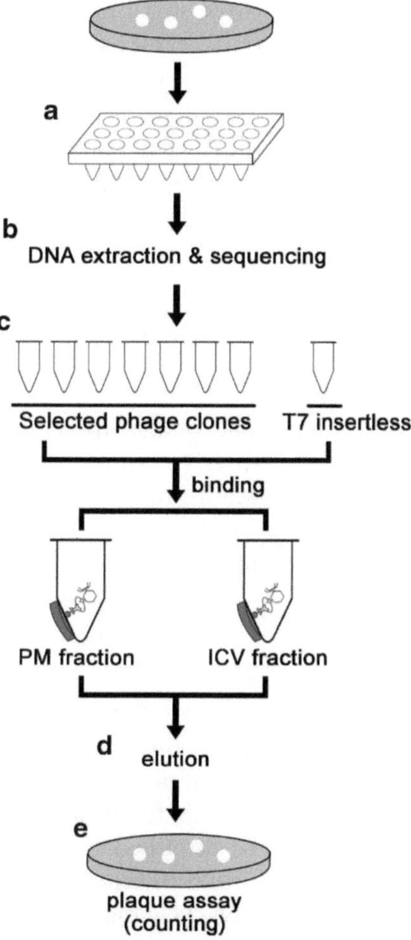

Fig. 2 Validation of the selected phage clones to the AQP2-expressing PM and/or ICV fractions. (**a**) After three rounds of panning, 80 plaques from the PM fractions and 80 plaques from the ICV fractions are randomly picked up from LB plates. (**b**) Each phage clone is sequenced and seven phage clones are chosen, which reveal amino acid sequences of high frequency. (**c**) High-affinity binding of these seven phage clones is examined by incubating each phage clone to the AQP2-immunoisolated fractions (both PM and ICV fractions), relative to the one by T7 insertless phage clone which does not display peptide ligands on the surface of the phage particle. (**d**, **e**) The phages bound to the complex are eluted (**d**) and subjected to a plaque assay for counting the number of phage clones (**e**). Thus, the high number of eluted each phage clone indicates the high affinity of each phage clone to the AQP2-immunoisolated fractions [7]

3.6 Labeling of Cells with Fluorescein-Conjugated Peptides

1. Seed IMCD cell suspension on human fibronectin-coated chamber slides. Feed the IMCD cells every 24 h. Grow them in hypertonic culture medium supplemented with 10 % FBS at 37 °C in 5 % CO_2, 95 % atmospheric air for 3 days, and then in FBS-free culture medium for an additional 1 day.
2. At day 5, fix IMCD cells with 2.5 % paraformaldehyde for 20 min at room temperature. After fixation, wash the cells

twice in PBS, and permeabilize with Triton X-100 at room temperature for 15 min.

3. Wash cells and block with 1 % BSA in 0.01 M PBS for 30 min and label with a fluorescein-labeled peptide (10 μM concentration, each) at 4 °C overnight. Synthetic peptides are conjugated to FITC. Control peptide (NSSVDK) conjugated to FITC is used for control experiment.
4. Wash cells and mount and carry out fluorescent microscopy.

4 Notes

1. The phage library is constructed according to the manufacturer's manual (T7Select® System Manual TB178; Novagen, Madison, WI). The library has a diversity of approximately 5×10^8 plaque-forming unit (pfu).
2. Peptides are synthesized by standard Fmoc method from a commercial company. The peptides are dissolved at 100 mM in DMSO and then slowly diluted with water until the concentration is 1 mM. Store in aliquots at −80 °C. Working solutions are prepared by dilution in water. Protect from light.
3. Generally, three to four rounds of screening are carried out to enrich phages. You may add additional two or three screening rounds for more enrichment. However, different phages can be amplified at different rates in growth cycle that follows each round of selection [11]. Thus increasing the number of selection cycle may lead to a selection of phages that grow fast rather than phages that obtain affinity.
4. It should be noted that there are limitations to identify specific protein candidates by library analysis using short peptide sequence. Because the peptide sequences we identified are short (nine amino acids consist of a randomly sequenced seven-amino acid residue between two cysteines) and did not match exactly the sequence of listed natural proteins, the probability of random match is high which is revealed by high E values in the BLAST. As per NCBI BLAST definition the p value of random matches is calculated as $p = 1 - e^{-E}$. Accordingly, E values higher than 15 yield a p value of 1, i.e., 100 % of random match.
5. The most frequently occurring peptide sequences are considered as promising ones. In most cases, you can find predominant sequences. In some cases, however, you may find several shared motifs of 3–4 amino acid length among different peptide sequences. For example, in our study seven phage clones, which demonstrated identical amino acid sequences of high frequency, are chosen from randomly selected 160 phage clones out of the enriched phage libraries for AQP2-expressing

PM fractions and ICV fractions [7]. Among them, phage clones displaying the sequences of CPKQRFWPC, CKRVTG RPC, and CKNMRSSAC constituted approximately 11 % of all the randomly selected phage clones (total 160). Fluorescein (or biotin) is usually attached at N-terminus of a peptide during synthesis.

Acknowledgements

This study was supported by National Research Foundation grant funded by the Ministry of Education, Science and Technology (MEST), Korea (2010-0008225 and 2010-0019393).

References

1. Knepper MA, Nielsen S, Chou CL, DiGiovanni SR (1994) Mechanism of vasopressin action in the renal collecting duct. Semin Nephrol 14: 302–321
2. Kwon TH, Nielsen J, Møller HB, Fenton RA, Nielsen S, Frøkiaer J (2009) Aquaporins in the kidney. Handb Exp Pharmacol 190:95–132
3. Nielsen S, Frokiaer J, Marples D, Kwon TH, Agre P, Knepper MA (2002) Aquaporins in the kidney: from molecules to medicine. Physiol Rev 82:205–244
4. Valenti G, Procino G, Tamma G, Carmosino M, Svelto M (2005) Minireview: aquaporin 2 trafficking. Endocrinology 146:5063–5070
5. Marples D, Knepper MA, Christensen EI, Nielsen S (1995) Redistribution of aquaporin-2 water channels induced by vasopressin in rat kidney inner medullary collecting duct. Am J Physiol 269:C655–C664
6. Rojek A, Fuchtbauer EM, Kwon TH, Frokiaer J, Nielsen S (2006) Severe urinary concentrating defect in renal collecting duct-selective AQP2 conditional-knockout mice. Proc Natl Acad Sci USA 103:6037–6042
7. Lee YJ, Choi HJ, Lim JS, Earm JH, Lee BH, Kim IS, Frokiaer J, Nielsen S, Kwon TH (2008) A novel method of ligand peptidomics to identify peptide ligands binding to AQP2-expressing plasma membranes and intracellular vesicles of rat kidney. Am J Physiol Renal Physiol 295:F300–F309
8. Lee SM, Lee EJ, Hong HY, Kwon MK, Kwon TH, Choi JY, Park RW, Kwon TG, Yoo ES, Yoon GS, Kim IS, Ruoslahti E, Lee BH (2007) Targeting bladder tumor cells in vivo and in the urine with a peptide identified by phage display. Mol Cancer Res 5:11–19
9. Hong HY, Lee HY, Kwak W, Yoo J, Na MH, So IS, Kwon TH, Park HS, Huh S, Oh GT, Kwon IC, Kim IS, Lee BH (2008) Phage display selection of peptides that home to atherosclerotic plaques: IL-4 receptor as a candidate target in atherosclerosis. J Cell Mol Med 12:2003–2014
10. Lee YJ, Song IK, Jang KJ, Nielsen J, Frokiaer J, Nielsen S, Kwon TH (2007) Increased AQP2 targeting in primary cultured IMCD cells in response to angiotensin II through AT1 receptor. Am J Physiol Renal Physiol 292:F340–F350
11. Azzazy HM, Highsmith WE Jr (2002) Phage display technology: clinical applications and recent innovations. Clin Biochem 35:425–445

Chapter 13

Protein Expression Profiling of Brain Tumor Tissue Using SELDI-MS

Carl Wibom

Abstract

Surface-enhanced laser desorption/ionization mass spectrometry (SELDI-MS) is an established, chip-based method for protein profiling, typically used for biomarker discovery. By combining retention chromatography and mass spectrometry on the same analytical platform, it allows for reliable analyses of small sample quantities in a high-throughput fashion. As such, it is a highly useful tool for a wide range of research fields. We have successfully applied it on brain tumor tissue samples to screen for differences in protein expression between invasive and noninvasive benign meningioma. This chapter lays out the details of the protocols we used, and can serve as a guide for protein expression profiling experiments on brain tumor tissue using SELDI-MS.

Key words Biomarkers, Brain tumor tissue, Surface-enhanced laser desorption/ionization (SELDI), Meningioma

1 Introduction

Investigations aimed at screening the low-molecular-weight proteome of complex biological samples are commonly performed by means of mass spectrometry (MS)-based approaches employing soft ionization. One of the main soft ionization techniques is matrix-assisted laser desorption/ionization (MALDI), which is employed both by conventional MALDI-MS as well as by its close relative, surface-enhanced laser desorption/ionization (SELDI)-MS. In conventional MALDI-MS, the samples generally need to be fractionated and purified separately before they are presented to the MS-system on a passive surface. In contrast, SELDI utilizes ProteinChip® arrays featuring chromatographically active surfaces (spots) that allow sample pretreatment to take place on the same chip that in turn is introduced to the MS-system [1]. The spots are designed to retain molecules with specific chemical properties by retention chromatography. During the SELDI process the crude sample is added to the spot surface and allowed to bind. Nonspecific

Helena Bäckvall and Janne Lehtiö (eds.), *The Low Molecular Weight Proteome: Methods and Protocols*, Methods in Molecular Biology, vol. 1023, DOI 10.1007/978-1-4614-7209-4_13,

interactions are subsequently removed by a washing procedure. The SELDI-MS-system is fitted with a ProteinChip® interface, thus enabling direct mass spectrometric characterization of the retained molecules. To permit for analyses of various proteome subsets with specific characteristics, there is a variety of array types with different chromatographic properties to choose from, including anion exchange (Q10), cation exchange (CM10), reverse-phase chromatography (H50), and immobilized metal affinity capture (IMAC). The chip-based sample fractionation and purification inherent to SELDI allow for high-throughput analyses with high intra-experimental analytical reproducibility. Furthermore, the high sensitivity of the linear time of flight (TOF) mass analyzers normally employed in SELDI-MS-systems ensures that relevant analyses can be performed on small sample volumes. This makes SELDI-MS a useful tool for proteomic investigations aimed at establishing differences between sample classes and detecting biomarkers, especially where sample quantity is a limiting factor.

We have applied the SELDI-MS approach to proteomic investigations of brain tumor tissue samples, collected from an experimental glioblastoma model [2] and from meningioma patients [3]. The latter was a retrospective study where we found protein expression patterns discriminating between invasive and noninvasive benign (WHO grade I) meningiomas. Meningioma is the most common primary brain or CNS tumor, and is most often classified as benign [4]. Benign meningiomas are typically characterized as noninvasive, have a relatively good prognosis, and may be cured by surgical resection. However, the prognosis depends largely on tumor location, where surgically accessible convexity tumors often have a better outcome than their skull base counterparts [5, 6]. Over time, inoperable tumors or tumor remnants may progress towards a more malignant phenotype [6, 7]. At present, there is no means to predict if an incompletely resected tumor will develop an invasive growth pattern. Establishment of prognostic biomarkers to this end has the potential to facilitate decisions concerning further treatment, such as whether radiotherapy should be given as an adjunct to surgery or at the time of recurrence [8, 9].

2 Materials

All plastic consumables, including tubes used for sample storage, should have low protein binding properties, such as polypropylene.

2.1 Tissue Sample Preparation

Both the homogenization buffers and the protein denaturing buffer listed below should be prepared in sufficient volumes to use on all samples in the study (*see* **Note 1**), then aliquoted, and stored at −20 °C.

1. Precellys®24 bead beating tissue homogenizer (Bertin Technologies, Saint-Quentin-en-Yvelines Cedex, France) or equivalent.
2. Lysing Matrix D tubes (Qbiogene, Montreal, Canada), i.e., 2 ml homogenization tubes with ceramic beads.
3. Homogenization buffer 1: 100 mM HEPES, pH 7.4, 100 mM NaCl, 0.5 % CHAPS including EDTA-free protease inhibitor cocktail.
4. Homogenization buffer 2: 5 M Guanidine–HCl, 50 mM Tris–HCl (pH 8.0), 0.5 % CHAPS including EDTA-free protease inhibitor cocktail.
5. Protein denaturing buffer: 8 M urea, 1 % CHAPS, PBS.
6. Liquid nitrogen.
7. BCA Protein Assay Reagent kit.

2.2 SELDI ProteinChip® Array Preparation

All buffers listed below should be prepared in large enough volumes to suffice throughout the study (*see* **Note 1**). Binding/wash buffers can normally be stored in room temperature for 3 months, although it is recommended that buffers are made fresh prior to the investigation. The properties of the binding/wash buffers for the various array types should be tested and optimized prior to a specific study (*see* **Notes 2** and **3**). A representative panel of standard, low-selectivity buffers is listed below.

1. ProteinChip® arrays (Bio-Rad Laboratories, Inc., Hercules, CA, USA).
2. Acetonitrile (ACN), HPLC grade.
3. Trifluoroacetic acid (TFA), HPLC grade.
4. 1 mM HEPES, pH 7.2.
5. H50 binding/wash buffer: 10 % ACN, 0.1 % TFA.
6. IMAC binding/wash buffer: 100 mM PO_4, pH 7.5, 0.5 M NaCl.
7. CM10 binding/wash buffer: 100 mM NaAc, pH 4.
8. Q10 binding/wash buffer: 100 mM Tris–HCl, pH 9.
9. Ion solution: 100 mM $CuSO_4$ (*see* **Note 2** and Table 1).
10. Matrix: Sinapinic acid (SPA)/alpha-cyano-4-hydroxycinnamic acid (CHCA), 5 mg/tube (*see* **Note 2**).
11. All-in-one protein standard/All-in-one peptide standard (Bio-Rad Laboratories, Inc.).
12. Bioprocessor (Bio-Rad Laboratories, Inc.).
13. Aluminum plate seals.
14. MicroMix5 shaker (Diagnostic Products Corporation, Los Angeles, CA, USA), or equivalent.

Table 1
Binding/wash buffer characteristics

	Parameter	Range[a]
H50	ACN	5–50 %
	TFA	0.1–1 %
	NaCl	0–0.2 M
IMAC	Metal ion	Cu/Zn[b]
	pH	4–8
	NaCl	0.5–1 M
	Imidazole	0–10 mM
CM10	Buffer salt	10–100 mM
	pH	4–7
Q10	Buffer salt	10–100 mM
	pH	7–9

[a]The listed ranges represent suggestions, not absolute values
[b]Other ion solutions can be used as well; see the manufacturer's instructions

15. Biomek 2000 Laboratory Automation Workstation robot (Beckman Coulter, Inc., Fullerton, CA, USA), or equivalent (optional, *see* **Note 4**).
16. Thermomixer Compact (Eppendorf, Hamburg, Germany), or equivalent (optional, *see* **Note 5**).
17. ProteinChip® reader, series PCS4000 (Bio-Rad Laboratories, Inc.).

3 Methods

When performing proteomic screening studies utilizing a label-free MS-approach, such as described here, it is critical that bias is not introduced to the collected data [10, 11]. Avoidance thereof can be ascertained by rigorous standardization and randomization of each sample handling and analysis step, including sample collection, freezing, storage, preparation, and analysis. Typical pitfalls include separate handling of samples belonging to different sample classes, and having more than one person prepare the samples and thereby risking the introduction of confounding variation. Furthermore, to facilitate the evaluation of analytical reproducibility it is suggested that multiple analytical replicates of a quality control (QC) sample are analyzed in parallel. By preparing numerous aliquots of the QC sample, storing them properly (usually at −80 °C), and consistently employing the same QC sample, it is possible to monitor the MS-system performance over time.

3.1 Tissue Sample Preparation

This protocol assumes the use of Precellys®24 or equivalent equipment for tissue homogenization. With slight adaptations the protocol works equally well with a manually operated Dounce Tissue Grinder (*see* **Note 6**). Unless otherwise stated, all sample handling steps should be performed on ice. To avoid introducing bias to the investigation, all samples must be prepared in a random order.

1. Weigh the frozen tissue sample and place it in a 2 ml homogenization tube with small ceramic beads. Add 1 ml homogenization buffer 1 per 100 mg tissue and let the sample thaw on ice. Repeat this step for each sample to be homogenized at once (*see* **Note 7**).
2. Place the homogenization tubes in the Precellys®24 tissue homogenizer and run three times for 30 s at 6,500 rpm (*see* **Note 8**).
3. Incubate the sample homogenate on ice for 30 min. Use the incubation time to transfer the sample homogenate to a new tube without beads.
4. Vortex the sample briefly and centrifuge it at 20,800 × *g* for 20 min at 4 °C.
5. Carefully transfer the supernatant to a separate tube and leave the pellet on ice.
6. Add two volumes protein denaturing buffer to the supernatant and incubate on a shaker for 30 min at 4 °C.
7. While the supernatant fraction is incubating (**step 6**), transfer the pellet fraction to a new 2 ml homogenization tube, add 1 ml homogenization buffer 2 per 100 mg tissue, and then rehomogenize the pellet fraction with the Precellys®24, three times for 30 s at 6,500 rpm (*see* **Note 8**).
8. Incubate the pellet homogenate on ice for 3 h. During this time, transfer the pellet homogenate to a new tube, without beads.
9. While the pellet homogenate is incubating on ice (**step 8**), aliquot the supernatant fraction (from **step 6**) into appropriate volumes (e.g., 50 μl), and snap freeze the aliquots in liquid nitrogen. Store at −80 °C until further analysis.
10. Repeat **step 4** for the pellet homogenate.
11. Carefully transfer the supernatant to a separate tube. Then aliquot the sample into appropriate volumes (e.g., 50 μl) and snap freeze the aliquots in liquid nitrogen. Store at −80 °C until further analysis.
12. Use one aliquot from each sample and fraction (supernatant/pellet) to determine the total protein concentration (*see* **Note 9**) by means of the BCA protein assay (see the manufacturer's instructions).

3.2 SELDI ProteinChip® Array Preparation

In this protocol, following each addition to the arrays (except matrix addition), the bioprocessor should be sealed with an aluminum seal and centrifuged at 80×*g* for 1 min (to ensure contact between the spot surface and the added fluid). Furthermore, all array incubations should take place on a MicroMix5 shaker, set to program 5 and amplitude 20. The bioprocessor is emptied between additions by turning it upside down over a waste container, and then hammering it repeatedly on paper towels spread out on the bench top. Throughout the protocol, polymer-free polypropylene pipette tips and non-latex gloves should be used.

1. Prepare an analysis scheme, where each analytical replicate of each sample is assigned a specific well in a 96-well plate. Make sure that samples belonging to various sample groups are thoroughly intermingled on the plate and that multiple replicates of the QC sample are assigned to random wells (*see* **Note 10**).
2. Thaw the samples on ice. Then, dilute them in binding buffer to a final concentration of 150 μg/ml (*see* **Note 11**) in a v-bottom 96-well plate on ice, in accordance with the analysis scheme prepared in **step 1**. Seal the plate, then briefly shake it on the MicroMix5 table, and keep it on ice until analysis (*see* **Note 12**).
3. Assemble the ProteinChip® arrays in a bioprocessor and begin pretreatment (*see* **Note 4**). For CM10 and Q10 arrays skip to **step 4**. H50 arrays are pretreated twice with 50 μl 50 % ACN for 5 min. IMAC arrays are initially charged with metal ions by two additions of 50 μl of a 100 mM ion solution for 5 min. Unbound metal ions are subsequently removed by two washes with 100 μl 1 mM HEPES for 5 min.
4. Equilibrate the spots three times with 150 μl array type-specific binding buffer for 5 min.
5. Shake the sample plate briefly on the MicroMix5 shaker, add 100 μl of the diluted samples (i.e., 15 μg protein/spot, *see* **Note 11**) to the arrays, and incubate for 1 h at room temperature.
6. While the samples are incubating on the ProteinChip® arrays (**step 5**), prepare a saturated matrix solution by adding 160 μl ACN:0.1 % TFA (1:1 by volume) to an amber matrix tube containing 5 mg CHCA or SPA and vortex the mixture for 15 min (*see* **Note 5**). Thereafter, centrifuge the solution at 15,300×*g* for 5 min. Carefully collect the supernatant and dilute it in an equal volume ACN:0.1 % TFA to achieve a 50 % saturated matrix solution, diluted in 50 % ACN and 0.5 % TFA. Protect the solution from light, either by using an amber tube or by wrapping the tube in aluminum foil.
7. Remove unbound sample and contaminants from the spots by three additions of 150 μl washing buffer (*see* **Note 3**) for 5 min. Wash thereafter twice with 150 μl 1 mM HEPES for 1 min.

8. Dismantle the bioprocessor, but leave the arrays in the bioprocessor frame and allow them to air-dry for 15 min in room temperature (*see* **Note 13**).
9. Carefully spot two volumes of 1 μl 50 % saturated matrix solution to the arrays, 3 min apart. Let the arrays air-dry, and store them wrapped in aluminum foil to protect them from light until MS-analyses (*see* **Note 14**).
10. Use the All-in-one protein or All-in-one peptide standard to prepare an array according to the manufacturer's specifications, to use for external calibration of the MS-system.

3.3 Data Acquisition

1. Load the arrays into the ProteinChip® reader system and optimize the system settings (*see* **Note 15**). Once the system settings have been selected, all samples that will ultimately be compared against each other must be analyzed using the same settings.
2. Calibrate the system by reading the standard peptide or protein mixture array, employing the same system settings that will be used for reading the sample arrays, i.e., those found best suited in **step 1**. If different settings are used for different mass ranges, the system should be calibrated separately for each mass range.
3. Construct automatic protocols for the analyses of each spot. Then read all spots in the mass ranges of interest.
4. Analyses of the acquired data can be performed using the vendor's supplied software or the data can be exported for further analyses, advantageously performed by an expert in multivariate data analysis.

4 Notes

1. To minimize the risk of introducing bias to the data, avoid using buffers from different batches within the same study.
2. Employing SELDI-MS to screen for proteomic differences between sample classes is often performed without a preconceived hypothesis regarding which proteome subset potentially harbors the putative discriminating features of interest. In other words, one normally does not know beforehand what experimental conditions to use in the study. As an initial step, it is therefore recommended that a small-scale screening is performed to elucidate what the optimal conditions are. Such a screen can include just a few samples from each of the various sample classes, and is ideally set up by factorial design of experiments, where one varies several experimental conditions simultaneously according to a particular design [12]. The main

parameters to optimize are selection of matrix molecule (SPA or CHCA) and the saturation thereof (50–100 % for SPA and 20–100 % for CHCA) as well as selection of chip type and the characteristics of the corresponding binding buffer. Table 1 lists buffer parameters to optimize and suggests a reasonable range for each.

3. The wash buffer is generally the same as the binding buffer, but one may experiment with different characteristics of the various buffers to affect the selectivity. To increase selectivity on (1) H50 arrays, increase the ACN content (5–50 %); (2) IMAC arrays, add imidazole to the buffer (5–10 mM); (3) CM10 arrays, increase the pH; and (4) Q10 arrays, decrease the pH.
4. The ProteinChip® arrays can be prepared manually, using a regular multichannel pipette. However, to enhance the analytical reproducibility over time, a pipetting robot, such as the Biomek 2000 Laboratory Automation Workstation, can be used for all ProteinChip® array preparation steps, including matrix addition. In any case, make sure that the pipetting tool is highly accurate and well calibrated.
5. Matrix solubility may be affected by fluctuations in room temperature, and may ultimately impact the readout. To minimize this effect and improve the reproducibility over time, use a Thermomixer Compact, or equivalent equipment, set to 25 °C and 1,400 rpm.
6. Using a Dounce Tissue Grinder for tissue homogenization is a time-consuming and laborious approach that essentially requires each sample to be prepared separately. If this approach is to be used, start with preparing just one sample at a time. Eventually it is possible to adapt the workflow to work with multiple samples. It is important that the glass equipment is thoroughly washed between handling different tissue samples, both with the lysis buffer and with water, to avoid cross contamination.
7. Although Precellys®24 has the capacity to handle 24 samples at a time, it is generally not practical to work with more than about six samples simultaneously.
8. The homogenization buffer contains the detergent CHAPS. After homogenization the sample usually appears to consist mainly of lather. This is to be expected. The lather will start to settle during the following incubation on ice. To handle the lathery sample, it may be more convenient to use a Pasteur pipette than an automatic pipette.
9. If there are too few aliquots of a sample to spend one on determining protein concentration, then estimate the concentration of that particular sample based on the measured concentrations

of the remaining samples, taking the weight of the tissue sample and buffer volumes used into account.

10. If more than one 96-well plate is used (i.e., more than 12 ProteinChip® arrays), then randomly distribute the samples (including replicates of the QC sample) over all plates.
11. The sample concentration suggested here (150 μg/ml) has proved to work well together with this protocol for all array types. However, for each individual study it is generally recommended that a separate experiment is performed beforehand, to determine the optimal sample quantity to add to each spot, normally judging by peak count and reproducibility. Keep in mind that the protein capacity varies between array types.
12. **Steps 1–2** may be performed in parallel with **steps 3–6**.
13. If multiple bioprocessors are used in the experiment, make sure that the air drying time and the interval between the first and second addition of matrix are consistent between bioprocessors. This may otherwise have a negative impact on reproducibility [13].
14. It is possible to read the arrays several weeks after preparation. It is however strongly recommended they are read shortly thereafter.
15. Although the SELDI process is relatively straightforward, the system optimization step normally requires some experience. Initially it is essential to establish a feeling for the range within which each parameter should be set. Thereafter, the settings should be systematically altered within these limits in search for a combination yielding a high-quality output signal. It is important to bear in mind that the signal intensity may vary dramatically between different pixels on the same spot, an occurrence known as the "sweet spot" phenomenon. Optimization can be performed by generating spectra through manually varying the parameters and selecting the pixels to read on each spot. This allows for high shot-to-shot consistency in terms of signal intensity, given that the "sweet spot" phenomenon is continuously accounted for. Alternatively, optimization can be based on a number of automatic protocols with systematically different settings, constructed to use pixels covering the entire spot surface. By studying how the output depends on the system settings it is possible to determine what the most suitable settings are. Considerations regarding some of the key system parameters to optimize are listed below, and Fig. 1 illustrates some of the spectral features that should be taken into account in this process. Given that sample quantity allows, optimization may advantageously be performed on a separate array, solely devoted for the purpose. If this is not an option, optimization can be based on a few selected spots of

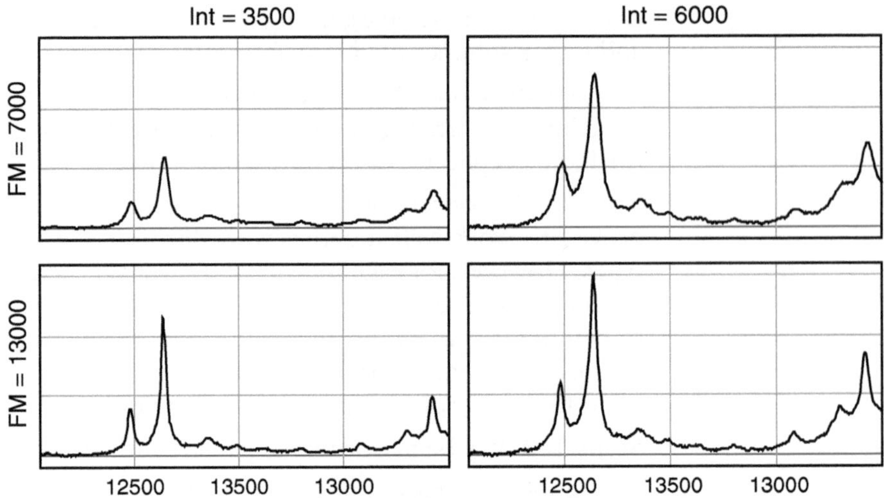

Fig. 1 Spectral features affected by MS-system settings. SELDI-MS spectra collected from the same spot, using different laser intensity (Int) and focus mass (FM) settings. These settings constitute two of the main MS-system parameters to optimize. The result of altering the FM from 7,000 Da (*upper panels*) to 13,000 Da (*lower panels*) is illustrated by the increased resolution for peaks around 13 kDa (the effect is opposite around 7 kDa). Furthermore, higher laser intensities (Int) generally generate a stronger output signal, as made obvious when comparing the *left-hand panels* (Int = 3,500) with the *right-hand panels* (Int = 6,000). Too high laser intensity may however negatively affect the resolution, in terms of valley depth between adjacent peaks, which is here particularly evident for the valley at 12.6 kDa

the arrays included in the study. In this case, care should be taken to avoid using the same spot pixels for optimization as will later be used for the actual analysis, to avoid significantly affecting the final readout.

Laser intensity (*Int*). This setting is a measure of the laser energy delivered with each shot. The intensity of the output signal is generally increased with increased laser intensity. At the same time, the spectral resolution may be reduced as the depth of the valleys between adjacent peaks is diminished (Fig. 1). The laser intensity should be set high enough to allow for detection of the smaller peaks, though low enough to keep the tallest peaks within measureable limits and with the adverse effects on resolution in mind.

Focus mass (*FM*). The SELDI-MS instrument normally utilizes time-lag-focusing. This allows the operator to define an *m/z*-value to focus the reading at, i.e., the spectral resolution will peak at this specific *m/z*-value and stepwise decrease with increased distance from this point. It is important to find an FM setting that produces sufficient resolution across the mass range of interest. To this end, it can be worthwhile dividing the analysis into smaller mass ranges and apply separate FM settings for each. Moreover, specific spectral features should also be taken into consideration. For instance, consider a situation where the mass range to analyze is 2–10 kDa and where there is a cluster of overlapping peaks between 7 and 8 kDa. Here, a reasonable FM setting would be 6,000 Da (in the center of the range). It may, however, prove beneficial to attempt and resolve the peak cluster by employing an FM setting around 7,500 Da, though this will be at the expense of a reduced resolution at the low end of the spectrum.

Matrix attenuation. The MS-system can deflect all molecules with an m/z-value lower than a specific cutoff (up to 10 kDa), and thereby prevent them from reaching the detector. The cut-off value is set by the matrix attenuation parameter, normally to 1,000–2,000 Da for the SPA matrix and to 500 Da for the CHCA matrix, to deflect the matrix molecules that otherwise produce a matrix-noise at the low end of the spectrum.

Sampling rate. This is a measure of the frequency by which the MS-system collects data. Higher frequencies allow for higher resolution, though they produce larger data files that take longer to process. 200–400 MHz is normally sufficient for profiling experiments; however 800 MHz may provide higher resolution in the low mass range.

Acknowledgment

The author thanks Dr. Mattias Forsell for critically reading the manuscript. Lina Mörén and Charlotte Nordström are gratefully acknowledged for their technical assistance. This work was in part funded by Lion's Cancer research foundation of northern Sweden.

References

1. Issaq HJ, Conrads TP, Prieto DA, Tirumalai R, Veenstra TD (2003) SELDI-TOF MS for diagnostic proteomics. Anal Chem 75:148A–155A
2. Wibom C, Pettersson F, Sjöström M, Henriksson R, Johansson M, Bergenheim AT (2006) Protein expression in experimental malignant glioma varies over time and is altered by radiotherapy treatment. Br J Cancer 94:1853–1863
3. Wibom C, Moren L, Aarhus M, Knappskog PM, Lund-Johansen M, Antti H, Bergenheim AT (2009) Proteomic profiles differ between bone invasive and noninvasive benign meningiomas of fibrous and meningothelial subtype. J Neurooncol 94:321–331
4. CBTRUS (2009) CBTRUS 2009 Statistical report: primary brain and central nervous system tumors diagnosed in the United States in 2004–2005, Central Brain Tumor Registry of the United States
5. Condra KS, Buatti JM, Mendenhall WM, Friedman WA, Marcus RB Jr, Rhoton AL (1997) Benign meningiomas: primary treatment selection affects survival. Int J Radiat Oncol Biol Phys 39:427–436
6. Lamszus K (2004) Meningioma pathology, genetics, and biology. J Neuropathol Exp Neurol 63:275–286
7. Bikmaz K, Mrak R, Al-Mefty O (2007) Management of bone-invasive, hyperostotic sphenoid wing meningiomas. J Neurosurg 107:905–912
8. Campbell BA, Jhamb A, Maguire JA, Toyota B, Ma R (2009) Meningiomas in 2009: controversies and future challenges. Am J Clin Oncol 32:73–85

9. Commins DL, Atkinson RD, Burnett ME (2007) Review of meningioma histopathology. Neurosurg Focus 23:E3
10. Mischak H, Apweiler R, Banks RE, Conaway M, Coon J, Dominiczak A, Ehrich JHH, Fliser D, Girolami M, Hermjakob H, Hochstrasser D, Jankowski J, Julian BA, Kolch W, Massy ZA, Neusuess C, Novak J, Peter K, Rossing K, Schanstra J, Semmes OJ, Theodorescu D, Thongboonkerd V, Weissinger EM, Van Eyk JE, Yamamoto T (2007) Clinical proteomics: a need to define the field and to begin to set adequate standards. Proteomics Clin Appl 1:148–156
11. Ransohoff DF (2005) Bias as a threat to the validity of cancer molecular-marker research. Nat Rev Cancer 5:142–149
12. Forshed J, Pernemalm M, Tan CS, Lindberg M, Kanter L, Pawitan Y, Lewensohn R, Stenke L, Lehtio J (2008) Proteomic data analysis workflow for discovery of candidate biomarker peaks predictive of clinical outcome for patients with acute myeloid leukemia. J Proteome Res 7:2332–2341
13. Aivado M, Spentzos D, Alterovitz G, Otu HH, Grall F, Giagounidis AAN, Wells M, Cho JY, Germing U, Czibere A, Prall WC, Porter C, Ramoni MF, Libermann TA (2005) Optimization and evaluation of surface-enhanced laser desorption/ionization time-of-flight mass spectrometry (SELDI-TOF MS) with reversed-phase protein arrays for protein profiling. Clin Chem Lab Med 43:133–140

Chapter 14

Comprehensive Analysis of MHC Ligands in Clinical Material by Immunoaffinity-Mass Spectrometry

Kie Kasuga

Abstract

Major histocompatibility complexes (MHC) are expressed on antigen-presenting cells (APC) that display peptide antigens. This is a crucial step to activate a T-cell response. Since immunogenic ligand of MHC is closely related with autoimmunity, inflammatory diseases, and cancer, comprehensive analysis of MHC ligands (the so-called *Ligandome*) is essential to unveil disease pathogenesis. Recently, immunotherapies such as vaccination have been focused on as new therapies of cancer, HIV, and infectious diseases. Therefore, the importance of comprehensive analysis of MHC ligands is increasing. Mass spectrometry has been the core technology of ligand identification since the 1990s. The sensitivity of mass spectrometers has been improved dramatically in recent years; thus, it enables to identify MHC ligands in clinical materials. This chapter lays out the workflow of MHC ligand identification in clinical materials, especially human bronchoalveolar (BAL) cells. MHC-ligand complexes are enriched by immunoaffinity extraction and captured ligand peptides are identified by LC-MS/MS. MHC class II ligand in BAL cells is described in this text; however, this approach is applicable to MHC class I and other clinical materials such as tissues.

Key words Major histocompatibility complexes (MHC), Immunoaffinity extraction, Ligand, Bronchoalveolar lavage (BAL), Comprehensive analysis

1 Introduction

Major histocompatibility complex (MHC) molecules, also referred as human leukocyte antigen (HLA), present peptides to be recognized by T-cell receptors. Two classes of MHC molecules are identified as class I and class II. Class I molecules consist of a membrane-inserted heavy chain (45,000 MW), and a non-covalently attached light chain (12,000 MW). Class II molecules are heterodimers and consist of two chains alpha and beta of similar size (30,000 MW) [1]. As shown in Fig. 1, MHC-I and -II are associated with different types of peptide ligands. MHC-I molecules bind 8–12 amino acids (AA) length of peptides, which are produced endogenously or degraded by the proteasome and other proteases in the cytoplasm and endoplasmic reticulum [2]. By contrast,

Helena Bäckvall and Janne Lehtiö (eds.), *The Low Molecular Weight Proteome: Methods and Protocols*, Methods in Molecular Biology, vol. 1023, DOI 10.1007/978-1-4614-7209-4_14, © Springer Science+Business Media New York 2013

MHC I

PDB. 1HSA

- Expressed on nucleated cells (epithelial cells,leukocytes, mesenchymal cells)
- Virus-antigen, Cancer-antigen (endogenous protein)
- Ligand peptides: 8 to 12 AA
- Cleavage by cytosolic proteasome (ex. 20S proteasome) and other proteases
- Site of Peptide loading: ER

MHC II

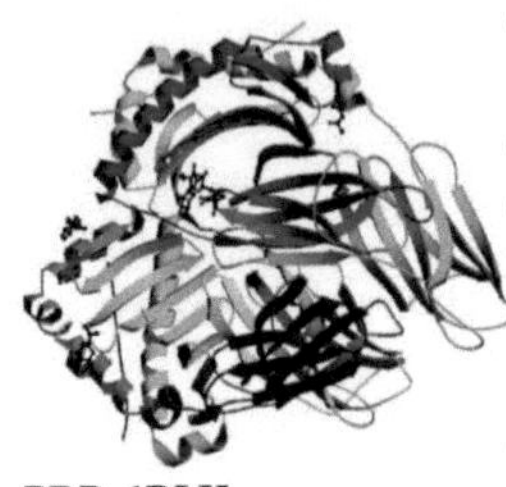

PDB. 1DLH

- Expressed mainly on B-Cell Macrophage, DC.
- Exogenous antigen
- Ligand peptides: 9 to 25 AA (mainly 15-22)
- Cleavage by various enzymes, such as cathepsin
- Site of Peptide loading: Specialized vesicular compartment

Fig. 1 Structures of MHC class I and class II with peptide presentation

MHC-II molecules bind 9–25 AA length of peptides, which are derived from exogenous, transmembrane, and cytosolic proteins. MHC-II ligand peptides are produced by various proteases originated from a lysosomal compartment [3, 4]. MHC-I-ligand complexes are recognized by $CD8^+$ T cells (cytotoxic T cells) and MHC-II-ligand complexes are recognized by $CD4^+$ T cells. Since recognition of peptides, which are derived from disease-associated or tumor-specific proteins, triggers a T-cell-mediated immune response, comprehensive analysis of MHC ligands is essential to unveil the mechanism of pathogenesis [5]. In early 1990s, Hunt DF et al. successfully identified MHC-II ligand peptide sequences by micro capillary-HPLC-MS/MS [6]. Since then, numerous number of studies have been reported using mass spectrometry. Recently we also presented our workflow of MHC-II ligand identification in clinical materials [7]. As shown in Fig. 2, MHC-ligand complexes are enriched from clinical samples. Followed by isolation of peptide ligands from MHC-ligand complexes, peptide mixtures are analyzed by nano-LC-MS/MS. In recent years, the sensitivity of mass spectrometers has been improved dramatically; thus, comprehensive analysis of MHC ligands, especially in clinical materials, has been available.

This chapter describes the workflow of MHC-II ligand identification from clinical materials, for instance, bronchoalveolar

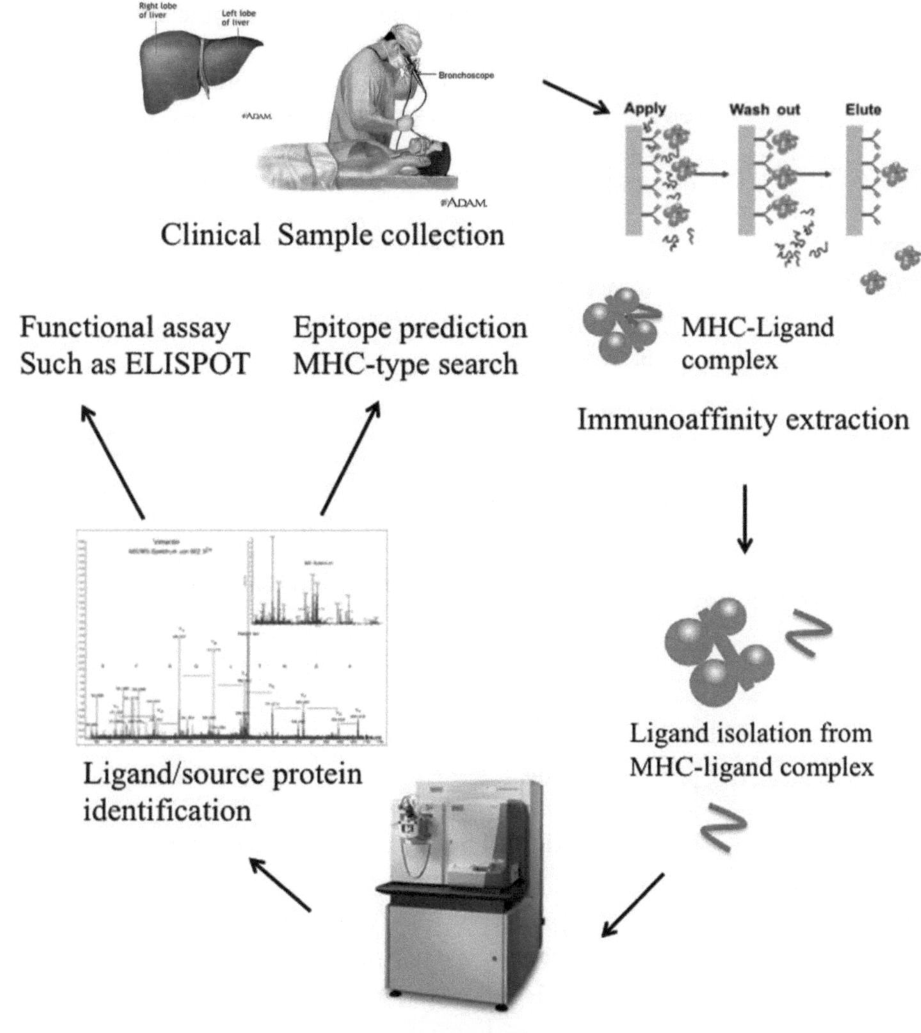

Fig. 2 Workflow: Comprehensive analysis of MHC ligands in clinical samples

lavage (BAL) cells. This method is applicable for MHC-I ligand identification, if MHC-I-specific antibodies are used. The biological validation, for instance, functional assay is beyond the scope of this chapter. In general, enzyme-linked immune spot assay (ELISPOT) is the popular technique to validate identified ligands. The principle and details of assay are described in [8].

2 Materials

Unless stated otherwise, solutions and mobile phase should be prepared in water that has a resistivity of 18.2 MΩ cm, such as MilliQ water. This standard is referred to as "water" in this text. All plastic consumables, including tubes used for sample storage, should have low protein absorption properties. Hydrophilic coating such as ProteoSave™ SS is a good option (*see* **Note 1**). All chemicals are preferably of the highest grade and buffers are sterile-filtered prior to using the column (*see* **Note 2**). Human BAL cells are selected as an example of clinical materials in this chapter.

2.1 Human Bronchoalveolar Collection

1. Phosphate-buffered saline (PBS, pH 7.4).
2. Anesthesia [for example, Morfin-skopolamin® (Meda) and Xylocain® (Astra Zeneca)].
3. Fiber-optic bronchoscope.
4. 50 ml disposal syringe.
5. Centrifuge (4 °C available).
6. Darcon net (Millipore).
7. Measuring cylinder.
8. Protease Inhibitor Cocktail (Sigma P8340, 100 times diluted with PBS, store at 4 °C).
9. Freeze medium (protease inhibitor:DMSO:RPMI medium = 10:90:900).
10. Bürker chamber.
11. Trypan blue.

2.2 Immobilization of Antibody or Isotype Control (IgG)

1. CNBr-activated Sepharose 4B (*see* **Note 3**).
2. 1 mM HCl.
3. Anti-human HLA-DR (MHC class II allele) monoclonal antibody (*see* **Note 4**) We use HB 55 (L243) (custom made by MABTECH AB, Sweden).
4. Isotype control (IgG), same animal species with antibody (*see* **Note 5**).
5. Coupling buffer: 0.5 M NaCl and 0.1 M $NaHCO_3$ in water, adjust to pH 8.3 by NaOH: Storage at 4 °C.
6. Blocking solution: 0.2 M Glycine, pH 8.0: Storage at 4 °C.
7. End-over-end mixer.
8. Bradford reagent.
9. 96-well plate.
10. Plate reader for 96-well plate (595 nm is available).

2.3 Sample Preparation for Immunoaffinity Extraction

1. Self-stand cryo tube, such as Hydrologix Tubes (Molecular Bioproducts, 2 ml volume).
2. Magnetic stirrer bar (cylindrical, 8 × 3 mm).

3. Magnetic stirrer.
4. Lysis buffer: 0.6 % CHAPS (Sigma) in PBS, add protease inhibitor just before use (*see* **Notes 6** and **7**).
5. 2× Lysis buffer: 1.2 % CHAPS in PBS, add protease inhibitor just before use.
6. Ultra sonicator.
7. Syringe filter 0.2 μm (e.g., Acrodisc® Syringe Filter PALL Life Science) (*see* **Note 8**).

2.4 Immunoaffinity Extraction of MHC-II–Peptide Complex

1. Lysis buffer for equilibration: 0.6 % CHAPS (Sigma) in PBS, add protease inhibitor just before use.
2. PBS.
3. 10 and 0.2 % Trifluoroacetic acid (TFA) for elution.
4. End-over-end mixer.
5. Speed-vac.

2.5 Peptide Isolation from MHC-II-Peptide Complex

1. Solid-phase extraction cartridge (e.g., Strata X, 33u Polymeric Reversed Phase 30 mg/1 ml, Phenomenex, *see* **Note 9**).
2. Methanol (HPLC grade).
3. Acetonitrile (ACN) (HPLC grade).
4. Formic acid (LC-MS grade).

2.6 Sample Analysis by Nano-LC-MS/MS System

1. Nano-HPLC system that is capable of delivering low flow rates, equipped with temperature-regulated auto-sampler with a reproducible injector.
2. Guard column.
3. Column: Reversed-phase (C18) column: 3–5 μm particle size, 75 or 100 μm internal diameter, and 15 cm length column are preferable (e.g., 15 cm long C18 picofrit column, 100 μm internal diameter, 5 μm bead size, Nikkyo Technos Co., Tokyo, Japan).
4. HPLC-grade ACN.
5. Formic acid (FA).
6. Water.
7. Solvent A: 97 % Water, 3 % ACN, and 0.1 % FA (*see* **Note 10**).
8. Solvent B: 5 % Water, 95 % ACN, and 0.1 % FA (*see* **Note 10**).
9. Mass spectrometer: Since accurate peptide sequence analysis is essential for this work, high-resolution mass spectrometer, such as quadrupole-time-of-flight (Q-TOF), Orbitrap, and fourier-transform (FT) mass spectrometers are highly recommended to use.

3 Methods

Since some of the ligands are not abundant and harder to ionize under electrospray ionization (ESI), contamination should be avoided. For instance, gloves should be worn in all procedures. In addition, refrain from chatting when handling the samples; otherwise unwanted contamination such as saliva-derived proteins may be identified.

3.1 Human Sample Collection

Before starting a project that is handling human materials, the regional ethical review board should approve the research plan. All patients and/or healthy subjects will be recruited from the hospital (institute) and written informed consent should be obtained from all of the participants. Diagnosis of patient is established by several findings including clinical manifestations and symptoms. One of the methods of BAL collection is shown below [9, 10].

1. After pre-medication with morphine–hyoscine (Morfin-skopolamin®, Meda) intramuscular and topical application of lidocaine (Xylocain®, Astra-Zeneca) bronchoscopy is performed with a flexible fiber-optic bronchoscope.
2. BAL is performed by wedging the bronchoscope in one of the subsegments of the middle lobe. For BAL, five 50 ml aliquots of warmed PBS are instilled and aspirated.
3. The fluid is collected in an ice-cold silicone-treated bottle kept on ice. The bottle with BAL cells is transported to the laboratory immediately after collection for further handling.
4. BAL fluid is strained through a double layer of Darcon net (Millipore) and the volume of recovered fluid is measured. Cells are pelleted by centrifugation at $400 \times g$, at 4 °C, for 10 min and the supernatants are poured off.
5. The cell pellets are resuspended in RPMI 1640 (Sigma-Aldrich). Cells are counted in a Bürker chamber, and total cell viability is determined by trypan blue exclusion (*see* **Note 11**).
6. BAL cells are centrifuged and RPMI medium is discarded. BAL cells are suspended with freeze medium, and then stored at −80 °C until sample preparation (*see* **Note 12**).

3.2 Immunoaffinity Extraction

Immunoaffinity extraction is a well-known method to enrich specific molecules from complicated biological matrixes, such as tissues, cell lysates, and body fluids. While an antibody binds strongly to its specific antigen, other molecules may bind with weak affinity, or bind nonspecifically on the surface of the antibody. Since clinical materials contain abundant proteins such as plasma proteins, eliminating nonspecific binding of these abundant proteins to antibody

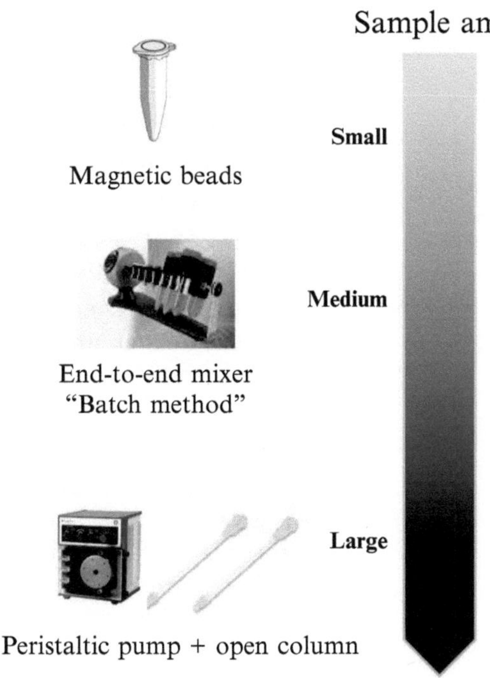

Fig. 3 Selection of affinity extraction systems

and solid-support is essential. Therefore, an optimization of washing conditions is a key factor of reliable immunoaffinity extractions. In MHC ligand analysis, first, MHC-ligand complexes are captured by an anti-MHC antibody. Following extraction of complexes, peptide ligands are isolated. The complex should be eluted without dissociating of ligand peptides from MHC molecules; thus, extraction and washing condition needs to be considered. Since conformational change of the antibody structure is observed to be pH dependent [11], acidic or basic elution is used for immunoaffinity extraction. Acidic elution is frequently used for MHC ligand identification [12]; however, successful result of basic elution is also reported [13]. The selection of solid-supports and scale of immunoaffinity extraction depend on the amount of samples (Fig. 3). Most importantly, the affinity of antibody against antigen is affected on wash and extraction procedures. Therefore, the condition of immunoaffinity extraction should be optimized for each of the antibodies that are used for the project. This chapter describes acidic elution with a batch method as an example.

3.2.1 Immobilization of Antibody

1. Activation of resin: Wash and swell 40 mg of cyanogen bromide-activated resin (CNBr) in cold 1 mM HCl for at least 30 min. The volume of the gel is ~1 ml (*see* **Note 13**).
2. Centrifuge at 50 × *g*, for 5 min, without brake. Discard supernatant (*see* **Note 14**).

3. Wash the resin with 5–10 volumes of water, and then equilibrate with coupling buffer. Immediately after equilibration, 10 mg of antibody or same amount of isotype control (IgG) will be transferred to Sepharose CNBr-4B (*see* **Note 15**).
4. Immobilize by mild end-over-end mixing at room temperature for 2 h or 4 °C overnight. Check the amount of non-bound antibodies or isotype control in supernatant by Bradford method to track immobilization efficiency (*see* **Note 16**).
5. Centrifuge at 50×*g*, for 5 min, without brake, and discard supernatant. Block unreacted groups with 0.2 M glycine (pH 8.0), with end-over-end mixing at room temperature for 2 h. Centrifuge at 50×*g*, for 5 min, without brake, and discard supernatant (*see* **Note 17**).
6. The antibody immobilized-resin is transferred to a vial, such as 5 ml cryogenic vial, and it is equilibrated with lysis buffer. If the resin is not used immediately, store the resin in PBS with 0.1 % NaN_3 at 4 °C.

3.2.2 Sample Preparation for Immunoaffinity Extraction

1. Thaw cell pellet in 1 volume of concentrated lysis buffer on ice (*see* **Note 18**). Let stand for ~10 min on ice. Gently pipette.
2. Transfer cells to a cryo tube and stir for 1 h at 4 °C.
3. Add 1 volume of lysis buffer and stir for another 1 h at 4 °C.
4. Cell lysate is further lysed by ultrasonication on ice. Sonicate 3× for 20 s with 20-s break in between sonication.
5. Add 1 volume of lysis buffer and stir for another 1 h at 4 °C.
6. Centrifuge at 1,500×*g* for 20 min at 4 °C. Collect the supernatant and discard pellet.
7. Supernatant: Ultracentrifuge for 1 h, 132,000×*g*, at 4 °C.
8. Filter supernatant by 0.2 μm filter.

3.2.3 Immunoaffinity Extraction of MHC Class II–Peptide Complex

1. Equilibrate with lysis buffer.
2. Load filtered cell lysate to antibody-immobilized or isotype control-immobilized gel (*see* **Note 19**).
3. Mix by end-over-end mixer overnight at 4 °C (*see* **Note 20**).
4. Centrifuge at 50×*g* for 5 min at 4 °C without brake. Discard supernatant.
5. Wash step 1: Add 3 beads volume of PBS, and mix by end-over-end mixer for 10 min at 4 °C. Centrifuge at 50×*g* for 5 min at 4 °C without brake. Discard Supernatant. Repeat this step five times.
6. Wash Step 2: Add 3 beads volume of water, mix by end-over-end mixer for 10 min at 4 °C Centrifuge 50×*g*, 5 min at 4 °C without brake. Discard supernatant. Repeat this step five times (*see* **Note 21**).

7. Add 100 μl of 10 % TFA to the gel, gently mix, and let stand for 5 min until the gel is shrinked. Add 100 μl of 0.2 % TFA and mix for 5 min. Centrifuge at 50 × *g* for 5 min without brake and collect supernatant.
8. Add 400 μl of 0.2 % TFA and mix for 5 min. Centrifuge at 50 × *g* for 5 min without brake and collect supernatant. Repeat this step five times.
9. Concentrate eluates by SpeedVac (*see* **Note 22**).

3.2.4 Peptide Isolation from MHC–Peptide Complex

1. Activation: Add 1 ml of MeOH to SPE cartridge (e.g., Strata X, *see* **Note 23**).
2. Equilibrate: Add 1.5 ml of water.
3. Eluates are resupended by water.
4. Apply eluates to SPE cartridge.
5. Wash: Add 2 ml of water.
6. Elution: Add 1 ml of ACN/1 % formic acid (7/3, v/v).
7. Eluates are concentrated by SpeedVac, and stored at −20 °C until analysis.

3.2.5 Peptide Identification by Nano-LC-MS System (Nano-LC-NSI-LTQ-Orbitrap)

Sensitivity of mass spectrometers has been improved; however, the identification of MHC ligands is still challenging. Proteomics studies routinely use trypsin, which cleaves after C-terminal of arginine and lysine to digest proteins to peptides. These peptides are left with at least two charges that favor ionization under ESI. On the other hand, MHC ligand peptides are generated by various proteases such as cathepsins. Therefore, MHC ligands quite often contain only a single charge, which makes it difficult to ionize properly under ESI. This creates a major sensitivity concern. Usually, only a couple of peptide sequences per protein are identified as MHC ligands. The improvement of technologies allows us to do "shot gun proteomics," for instance, protein identification based on a couple of peptides. Ligands and source protein identification are thus possible at the same time. However, peptides are frequently common between multiple proteins; therefore, manual inspection of data or BLAST search is recommended. In addition, high-quality fragment spectra (MS/MS) with high mass accuracy are essential to identify ligands. Taken together, sample preparation and handling should be prudent.

1. Sample is reconstituted with 10 μl of solvent A.
2. Set up the gradient system. For example, the curved gradient went from 2 % B up to 40 % B in 45 min, followed by a steep increase to 100 % B in 5 min (*see* **Note 24**).
3. C18 guard desalting column is set prior to a 15 cm long C18 picofrit column (100 μm internal diameter, 5 μm bead size, Nikkyo Technos Co., Tokyo, Japan) installed on to the nano electrospray ionization (NSI) source.

4. Parameters of ionization such as collision-induced dissociation (CID) depend on the instrument. For example, precursors were isolated with a 2 *m/z* width and dynamic exclusion was used with 60-s duration. We enabled "preview mode" for FTMS master scans, which proceeded at 30,000 resolution (profile mode). Data-dependent MS/MS (centroid mode) followed in two stages: firstly, the top five ions from the master scan were selected for CID (at 35 % energy) with detection in the ion trap (ITMS); and after, the same five ions underwent higher energy collision dissociation (HCD, at 45 % energy) with detection in the orbitrap (FTMS). The entire duty cycle lasted ~3.5 s. Detect single-charged peptides to multiple-charged peptides, since ligand peptides are created by proteases and degradation (*see* **Note 25**).
5. Raw data file is submitted to database search to identify peptide sequences and source proteins.

3.2.6 Ligand Identification by Database Search

There are two distinct approaches for peptide identification, such as database search and de novo sequencing. Database search is a statistical process. This is the popular peptide identification approach, and search in a given database to find target peptides. Results are cut off by false discovery rate (FDR). On the other hand, de novo sequencing is the analytical process that derives a peptide's amino acid sequence from its tandem mass spectrum (MS/MS) without the assistance of a sequence database. De novo sequencing enables to identify novel peptides which are not involved in a database; this is an advantage of this approach. We prefer database search since accuracy of peptide identification is prioritized. For de novo sequencing, several algorithms (software) are commercially available, for instance, PEAKS. (R) [14] (*see* **Note 26**).

1. Raw data (MS/MS spectra of MHC ligands) is submitted to database search, such as MASCOT or SEQUEST.
2. Set parameters. Enzyme parameter should be selected with "nonenzymatic."
3. Identified ligands are compared between Ab-immobilized gels and negative control (isotype-control-immobilized gel and gel itself). Ligand peptides obtained from isotype-control should be subtracted from the results. It would be nonspecific binding peptides to solid-support or surface of antibody.

3.2.7 Epitope Prediction by Bioinformatics Approach

Following ligand identification, further investigation of epitopes in ligands is critical. The bioinformatics approach is quite useful, and more than 15 algorithms can be freely available online as of May 2012. The list of algorithms and Web site is summarized in [15]. For T-cell epitopes, currently two distinct classes of algorithms are popular, for instance, MHC binding-motif-based algorithms and machine learning-based algorithms. One of the major motif-based algorithms is SYFEPITHI (http://www.syfpeithi.de/) [16], and machine learning-based algorithm is SVMHC (http://abi.inf.uni-tuebingen.de/Services/SVMHC/index_html) [17]. In addition to these algorithms, epitope-associated database, such as IEDB (The Immune Epitope Database and Analysis Resources: http://www.immuneepitope.org) is available.

With improvement of bioinformatics approach, more accurate prediction algorithms would be developed in the near future.

4 Notes

1. ProteoSave™ SS (Sumitomo Bakelite, Co Ltd.): Detailed information is as follows: http://www.sumibe.co.jp/english/product/s-bio/protein/proteo/index.html.
2. Same as other type of immunoaffinity extraction, contaminants in solution and buffer may disturb immobilization of antibodies as well as the interaction between an antibody and antigens. In case of column chromatography, contaminants may be the cause of clogging of a column.
3. A variety of solid-phase supports for immunoaffinity extraction are available. The selection would be made by the available clinical sample amount or anticipated expression amount of MHC molecules. Table 1 describes frequently used solid-supports for immunoaffinity extraction. In case of samples from tissue or certain amount of cells, CNBr-activated Sepharose 4B would be the first choice because of its capacity and reproducibility. CNBr-based immobilization is able to immobilize IgG under mild condition, and this is a critical factor. In addition, CNBr-Sepharose 4B has large capacity comparing with magnetic beads, and immobilization is a simple process. This support can be used both for open column chromatography and batch methods. If only small amount of sample is available, a scale down of chromatography is required. Batch method is economical; however, complete collection of eluates is quite difficult. Magnetic beads have an advantage of its low-dead volume, although it is relatively expensive (Fig. 3).
4. The selection of the antibody (Ab) in immunoaffinity extraction is crucial. The selection is based on the type of HLA alleles, monoclonal or polyclonal antibody, or antibody affinity. Polyclonal Abs are frequently used for immunoaffinity extraction, since it recognizes multiple epitopes; however, cross-reactivitiy is an issue. We use monoclonal Abs because antigen specificity is prioritized. The affinity of Abs may exert the condition of washing and elution procedures of immunoaffinity extraction. Therefore, method optimization for each Ab used for experiments may be required.

Table 1
Examples of solid-supports for immunoaffinity extraction

		Matrix	Particle size	Ligand	Binding capacity (/1 ml gel)
CNBr-sepharose 4B	GE Healthcare	4 % Agarose	90 μm	–	25–60 mg a-chymotrypsin
Affigel 10	Bio-Rad	*N*-hydroxysuccinimide esters of a delivatized cross-linked agarose gel beads with 10-atom spacer arm	–	–	30 mg of human γ globrin
Protein A mag sepharose	GE Healthcare	Paramagnetic, spherical, highly cross-linked agarose particles	37–100 μm	Protein A	8–17 mg human IgG
Protein A mag sepharose	GE Healthcare	Paramagnetic, spherical, highly cross-linked agarose particles	37–100 μm	Protein G	13–22 mg human IgG
Dynabeads®: Protein A	Invitrogen	Superparamagnetic matrix	–	Protein A	0.25 mg human IgG
Dynabeads®: Protein G	Invitrogen	Superparamagnetic matrix	–	Protein G	0.64 mg mouse IgG (7 μg human IgG/mg beads)

Information from Bio-Rad, GE Healthcare, and Invitrogen

5. Negative controls such as IgG (isotype control) and solid-support itself are strongly recommended to use as controls. Nonspecific binding peptides to antibody and resin should be determined.
6. The effective time of protease inhibitors is short. Therefore, protease inhibitors should be added immediately before use. For example, Complete Mini (Roche) is effective at 4 °C for 1–2 weeks. See the product instructions.
7. Nonionic detergents, such as Triton X-100 and Tween, are frequently used to solubilize proteins. These detergents affect the mass spectrometry analysis since polymer-like peaks derived from nonionic detergents masked peptide peaks. These detergents are quite difficult to remove from solutions.

   Although CHAPS also remains in the sample, it is eluted later than MHC ligand peptides. Therefore, CHAPS can be used for MHC ligand analysis by mass spectrometry.
8. Cell debris may cause column clogging. Filtering of samples will remove the debris to prevent clogging.
9. Alternatively, size exclusion filter unit, such as Amicon Ultra-4 (UFC800524, Millipore, 24/pk, 5K MWCO), can be used. Selection may be made by the amount of samples. Filter may show adsorption; thus, MHC ligands may be lost, in case.
10. To obtain a reproducible gradient effect, ACN should be added to both solvent A and B. Differences of solvent viscosity have effect on function of HPLC pump.
11. Upon objective of the project, cell differential counts should be conducted. The protocol is shown as follows: Smears for differential counts are prepared by cytocentrifugation at 50 × *g* for 3 min (Cytospin 2 Shandon; Southern Products Ltd.), whereupon cells are stained with May-Grünwald–Giemsa and approximately 500 cells are counted.
12. Fresh BAL cells are ideal for MHC ligand identification. However, it may be difficult to obtain clinical materials in good time. Alternatively, BAL cells can be stored in freeze medium at −80 °C. Since freeze and thaw cycle may cause dissociation of MHC-ligand complexes as well as degradation, this should be avoided.
13. These steps are necessary to remove the stabilizers, for example, lactose. Stabilizers will interfere with immobilization. The use of HCl preserves the activity of the reactive groups.
14. Brake of a centrifuge should not be used. The resin (beads) may be collapsed.
15. In general, antibodies are stored in solution with NaN_3. Sometime NaN_3 interferes with the coupling reaction for some of solid-supports (resin). If immobilization does not proceed

well, NaN_3 might be one of the reasons. The amount of antibodies that is used for immunoaffinity extraction is difficult to predict. It depends on the sample amount and expression levels of MHC molecules. In addition, antibodies may be immobilized to the solid-support with incorrect orientation, since immobilization is random with different sites of the antibodies. For instance, multiple point attachments of antibodies can induce steric hindrance to antigen binding. The Fab region of the antibodies, that is the antigen-binding region, has been used for coupling to the resin. To overcome this issue, site-directed immobilization, which is immobilization of Ab at Fc region, is used. If possible, use of extra amount of antibody for immobilization would be ideal for effective immunoaffinity extraction.

16. Unbound antibody level in solution is determined by Bradford method using 96-well plate (micro volume) assay.
17. Blocking of antibody-free surface is an important step to improve MHC ligand identification. If blocking is not completed, MHC molecules will be attached to the surface of solid-support. As a result, these MHC molecules might be lost without capturing. The principle of blocking is hydrolyzation of reactive groups on resins in basic solution. See the instruction of Sepharose by the company; sometime it shows an alternative blocking protocol. For example, block with 1 M ethanolamine for 2 h with end-over-end mixing, and wash with 0.1 M acetate buffer (pH 4.0) containing 0.5 M NaCl following wash with coupling buffer. Complete this wash cycle four to five times. In case of magnetic beads, some products are already made blocking.
18. In human BAL cells, the volume of 2×10^6 cells is approximately 0.2 ml based on our experiences.
19. In 5 ml cryogenic tube, cell lysate would be up to volume of 2 ml for effective agitation.
20. Mixing for 2 h at room temperature is an alternative way; however, 4 °C is recommended, since unwanted degradation may occur at room temperature.
21. This extensive washing step is required to reduce CHAPS in eluates. Also, it helps to reduce plasma-derived contamination, such as serum albumin-derived peptides, fibrin, etc.
22. Since TFA is not compatible with Strata X, eluates should be evaporated to remove TFA and redissolved with water.
23. These procedures are applicable to most of SPE cartridge. Resin that is used for Strata X is relatively tolerant to dryness; however, drying up of an SPE cartridge should be prevented.
24. Gradient time and ratio depend on complexity of sample and length of column.

25. If fragmentation (MS/MS), or quality of fragmentation is not enough, try to use different types of instruments or different fragmentation methods, such as electron transfer dissociation (ETD) [18].
26. In addition to PEAKS®, several de novo sequencing algorithms are available. However, most of these algorithms are designed for specific enzymatic digested peptides, such as trypsin. MHC ligands are randomly cleaved peptides (parameter as a nonenzymatic digestion). Also, mutated peptides are frequently found in virus, bacteria, and cancer; thus, these peptides are also possible ligand peptides. Since de novo sequencing is also useful to identify mutated peptides, the development or the improvement of algorithms that fit for MHC ligand identification would be highly appreciated.

Acknowledgments

Dr. Stefan Stevanović (Department of Immunology, Institute for Cell Biology, University of Tübingen, Tübingen, Germany) and Dr. Rui Mamede Branca (Department of Oncology-Pathology, Karolinska Institutet, Science for Life Laboratory Stockholm) are gratefully acknowledged for helpful discussion. The author is a VINNMER fellow supported via VINNMER Marie Curie Chair—a VINNOVA programme cofounded by Marie Curie Actions.

References

1. Klein J (1986) Natural History of the Major Histocompatibility Complex, 99th edn. Wiley, New York
2. Rammensee HG (1995) Chemistry of peptides associated with MHC class I and class II molecules. Curr Opin Immunol 7:85–96
3. Chicz RM, Urban RG, Gorga JC, Vignali DA, Lane WS, Strominger JL (1993) Specificity and promiscuity among naturally processed peptides bound to HLA-DR alleles. J Exp Med 178:27–47
4. Dongre AR, Kovats S, deRoos P, McCormack AL, Nakagawa T, Paharkova-Vatchkova V, Eng J, Caldwell H, Yates JR 3rd, Rudensky AY (2001) In vivo MHC class II presentation of cytosolic proteins revealed by rapid automated tandem mass spectrometry and functional analyses. Eur J Immunol 31:1485–1494
5. Banchereau J, Palucka AK (2005) Dendritic cells as therapeutic vaccines against cancer. Nat Rev Immunol 5:296–306
6. Hunt DF, Michel H, Dickinson TA, Shabanowitz J, Cox AL, Sakaguchi K, Appella E, Grey HM, Sette A (1992) Peptides presented to the immune system by the murine class II major histocompatibility complex molecule I-Ad. Science 256:1817–1820
7. Kasuga K, Branca RM, Wahlstrom J, Wheelock AM, Eklund A, Grunewald J, Lehtio J (2011) High sensitivity approach for identification of MHC Class II-bound peptides from bronchoalveolar lavage cells in sarcoidosis patients. Am J Respir Crit Care Med 183:A6404
8. Horlock C, Stott B, Dyson J, Ogg G, McPherson T, Jones L, Sewell AK, Wooldridge L, Cole DK, Stebbing J, Savage P (2009) ELISPOT and functional T cell analyses using

HLA mono-specific target cells. J Immunol Methods 350:150–160

9. Eklund A, Blaschke E (1986) Relationship between changed alveolar-capillary permeability and angiotensin converting enzyme activity in serum in sarcoidosis. Thorax 41:629–634

10. Wahlstrom J, Dengjel J, Winqvist O, Targoff I, Persson B, Duyar H, Rammensee HG, Eklund A, Weissert R, Grunewald J (2009) Autoimmune T cell responses to antigenic peptides presented by bronchoalveolar lavage cell HLA-DR molecules in sarcoidosis. Clin Immunol 133:353–363

11. Mano N, Abe K, Goto J (2006) Immunoaffinity extraction of a peptide modified by a small molecule. Anal Biochem 349:254–261

12. Friede T, Gnau V, Jung G, Keilholz W, Stevanovic S, Rammensee HG (1996) Natural ligand motifs of closely related HLA-DR4 molecules predict features of rheumatoid arthritis associated peptides. Biochim Biophys Acta 1316:85–101

13. Nag B, Passmore D, Deshpande SV, Clark BR (1992) In vitro maximum binding of antigenic peptides to murine MHC class II molecules does not always take place at the acidic pH of the in vivo endosomal compartment. J Immunol 148:369–372

14. Ma B, Zhang K, Hendrie C, Liang C, Li M, Doherty-Kirby A, Lajoie G (2003) PEAKS: powerful software for peptide de novo sequencing by tandem mass spectrometry. Rapid Commun Mass Spectrom 17:2337–2342

15. Chen P, Rayner S, Hu KH (2011) Advances of bioinformatics tools applied in virus epitopes prediction. Virol Sin 26:1–7

16. Rammensee H, Bachmann J, Emmerich NP, Bachor OA, Stevanovic S (1999) SYFPEITHI: database for MHC ligands and peptide motifs. Immunogenetics 50:213–219

17. Donnes P, Elofsson A (2002) Prediction of MHC class I binding peptides, using SVMHC. BMC Bioinformatics 3:25

18. Syka JE, Coon JJ, Schroeder MJ, Shabanowitz J, Hunt DF (2004) Peptide and protein sequence analysis by electron transfer dissociation mass spectrometry. Proc Natl Acad Sci USA 101(26):9528–9533

INDEX

Helena Bäckvall and Janne Lehtiö (eds.), *The Low Molecular Weight Proteome: Methods and Protocols*, Methods in Molecular Biology, vol. 1023, DOI 10.1007/978-1-4614-7209-4, © Springer Science+Business Media New York 2013

K

L

M

N

O

P

Q

R

MIX
Papier aus verantwortungsvollen Quellen
Paper from responsible sources
FSC® C105338

If you have any concerns about our products,
you can contact us on
ProductSafety@springernature.com

In case Publisher is established outside the EU,
the EU authorized representative is:
Springer Nature Customer Service Center GmbH
Europaplatz 3, 69115 Heidelberg, Germany

Printed by Libri Plureos GmbH
in Hamburg, Germany